P9-CKR-563

Evocative Objects

Evocative Objects

Things We Think With

edited by Sherry Turkle

 L.C.C.C. LIBRARY

The MIT Press   Cambridge, Massachusetts   London, England

© 2007 Massachusetts Institute of Technology

All rights reserved. No part of this book may be reproduced in any form by any electronic or mechanical means (including photocopying, recording, or information storage and retrieval) without permission in writing from the publisher.

MIT Press books may be purchased at special quantity discounts for business or sales promotional use. For information, please Email <special_sales@mitpress.mit.edu> or write to Special Sales Department, The MIT Press, 55 Hayward Street, Cambridge, MA 02142.

This book was set in Bookman Old Style, ITC Bookman, and Stymie by Graphic Composition, Inc., Athens, Georgia.

Printed and bound in the United States of America.

Library of Congress Cataloging-in-Publication Data

Evocative objects : things we think with / [edited by] Sherry Turkle ;
 theoretical essay and bibliographical essay by Sherry Turkle.
    p. cm.
  Includes bibliographical references and index.
  ISBN-13: 978-0-262-20168-1 (hardcover : alk. paper)
  1. Transitional objects (Psychology) I. Turkle, Sherry.
 BF175.5.T73E96 2007
 155.9'1—dc22
                                                        2006027966

10 9 8 7 6 5 4 3

To Charles Zimmerman

4/9/08   24.95

Contents

viii   Acknowledgments

3   Sherry Turkle | Introduction: The Things That Matter

OBJECTS OF DESIGN AND PLAY
12   Tod Machover | *My Cello*
22   Carol Strohecker | *Knots*
30   Susan Yee | *The Archive*
38   Mitchel Resnick | *Stars*
46   Howard Gardner | *Keyboards*

OBJECTS OF DISCIPLINE AND DESIRE
54   Eden Medina | *Ballet Slippers*
62   Joseph Cevetello | *The Elite Glucometer*
70   Matthew Belmonte | *The Yellow Raincoat*
76   Michelle Hlubinka | *The Datebook*
86   Annalee Newitz | *My Laptop*
92   Gail Wight | *Blue Cheer*

OBJECTS OF HISTORY AND EXCHANGE
102   Julian Beinart | *The Radio*
110   Irene Castle McLaughlin | *The Bracelet*
118   David Mitten | *The Axe Head*
126   Susan Spilecki | Dit Da Jow: *Bruise Wine*
136   Nathan Greenslit | *The Vacuum Cleaner*

OBJECTS OF TRANSITION AND PASSAGE
144   William J. Mitchell | *The Melbourne Train*
152   Judith Donath | *1964 Ford Falcon*
162   Trevor Pinch | *The Synthesizer*
170   Tracy Gleason | *Murray: The Stuffed Bunny*
178   David Mann | *The* World Book
184   Susan Rubin Suleiman | *The Silver Pin*

OBJECTS OF MOURNING AND MEMORY

194 Henry Jenkins | *Death-Defying Superheroes*

208 Stefan Helmreich | *The SX-70 Instant Camera*

216 Glorianna Davenport | *Salvaged Photographs*

224 Susan Pollak | *The Rolling Pin*

232 Caroline A. Jones | *The Painting in the Attic*

244 Olivia Dasté | *The Suitcase*

OBJECTS OF MEDITATION AND NEW VISION

252 Nancy Rosenblum | *Chinese Scholars' Rocks*

260 Susannah Mandel | *Apples*

270 Jeffrey Mifflin | *The Mummy*

278 Michael M. J. Fischer | *The Geoid*

286 Robert P. Crease | *Foucault's Pendulum*

296 Evelyn Fox Keller | *Slime Mold*

307 Sherry Turkle | *What Makes an Object Evocative?*

328 Notes

344 Selected Bibliography

364 Epigraph Sources

370 Illustration Credits

374 Index

Acknowledgments

This book began with a seminar series at the MIT Initiative on Technology and Self and became a way to capture the intellectual enthusiasms of that enterprise. I thank the Mitchell Kapor Foundation for its support of the Initiative as well as all participants in the Evocative Objects Seminar in the STS program over many years.

My research assistants Anita Chan, Olivia Dasté, and Kelly Gray worked closely with me on all aspects of making this volume a reality. Deborah Cantor-Adams's labors ensured consistency and clarity. I am grateful for Erin Hasley's combination of tenacity and perfect pitch in the design work for this volume. Lisa Liu's excellent transcription and editing of Initiative presentations helped many of the authors recall unscripted moments that improved the quality of their final essays.

Thanks are also due to Valerie Geary, Mark Kramer, Robert Prior, and Susan Silbey for their helpful comments on my introduction and concluding essay. Comments on the introduction by Michael Fischer, Howard Gardner, and Susan Suleiman are gratefully acknowledged. Funding from the Intel Corporation to pursue work on the complex qualities of objects and conversations with my research partner at Intel, Margaret Morris, have deepened my thinking.

This book has been a labor of love; I have lived with it for many years. I thank my daughter Rebecca for lighting up my life as I worked. Tellingly, she has resisted my recent suggestions that she tidy up her room by informing me that the stuff I want her to throw out are her evocative objects. I'm taking this as a good sign that the phrase is apt to catch on with others.

Sherry Turkle
Boston, Massachusetts
January 2007

Evocative Objects

INTRODUCTION: THE THINGS THAT MATTER

Sherry Turkle

I grew up hoping that objects would connect me to the world. As a child, I spent many weekends at my grandparents' apartment in Brooklyn. Space there was limited, and all of the family keepsakes—including my aunt's and my mother's books, trinkets, souvenirs, and photographs—were stored in a kitchen closet, set high, just below the ceiling. I could reach this cache only by standing on the kitchen table that I moved in front of the closet. This I had been given permission to do, and this is what I did, from age six to age thirteen or fourteen, over and over, weekend after weekend. I would climb onto the table in the kitchen and take down every book, every box. The rules were that I was allowed to look at anything in the closet, but I was always to put it back. The closet seemed to me of infinite dimensions, infinite depth.

Each object I found in the closet—every keychain, postcard, unpaired earring, high school textbook with its marginalia, some of it my mother's, some of it my aunt's—signaled a new understanding of who they were and what they might be interested in; every photograph of my mother on a date or at a dance became a clue to my possible identity. My biological father had been an absent figure since I was two. My mother had left him. We never spoke about him. It was taboo to raise the subject. I did not feel permitted to even think about the subject.

My aunt shared the small apartment with my grandmother and grandfather, and sometimes one of them would come into the kitchen to watch me at my investigations. At the time I didn't know what I was looking for. I think they did. I was looking, without awareness, for the one who was missing. I was looking for a trace of my father. But they had been there before me and gotten rid of any bits and pieces he might have left—an address book, a business card, a random note. Once I found a photograph of a man standing on a boardwalk with his face cut out of the picture. I never asked whose face it was; I knew. And I knew enough never to mention the photograph, for fear that it too would disappear. It was precious to me. The image had been attacked, but it contained so many missing puzzle pieces. What his hands looked like. That he wore lace-up shoes. That his pants were tweed.

If being attentive to the details of people's lives might be considered a vocation, mine was born in the smell and feel of the memory closet and its objects. That is where I found the musty books, photographs, corsages, and gloves that made me feel connected. That is where I determined that I would solve mysteries and that I would use objects as my clues.

Years from then, in the late 1960s, I studied in Paris, immersed in the intellectual world of French structuralism. While I was away, my grandparents moved out of their apartment, where the contents of the memory closet had been so safely contained. Much of the closet's contents were dispersed, sent to an organization that collected books to be read to the blind. Far away from home, I was distressed at the loss of the objects but somewhat comforted to realize that I now had a set of ideas for thinking about them. In Paris, I read the work of the anthropologist Claude Lévi-Strauss, who described *brico-lage* as a way of combining and recombining a closed set of materials to come up with new ideas.[1] Material things, for Lévi-Strauss, were goods-to-think-with and, following the pun in French, they were good-to-think-with as well. While in France, I realized that during my many hours with the memory closet I had done more than daydream ideas into old photographs. When I first met the notion of bricolage, it already seemed like a childhood friend.

Ideas about bricolage were presented to me in the cool, cognitive light of French intellectual life. But the objects I tried to combine and recombine as a child had been clues for tracing my lost father, an experience of bricolage with a high emotional intensity. So, from my first introduction to the idea in the late 1960s, I began to consider bricolage as a passionate practice.

We find it familiar to consider objects as useful or aesthetic, as necessities or vain indulgences. We are on less familiar ground when we consider objects as companions to our emotional lives or as provocations to thought. The notion of evocative objects brings together these two less familiar ideas, underscoring the inseparability of thought and feeling in our relationship to things. We think with the objects we love; we love the objects we think with.

In this collection of autobiographical essays, scientists, humanists, artists, and designers trace the power of objects in their lives, objects that connect them to ideas and to people. Some of the objects described in this book are natural: an apple. Some are artifacts: a train. Some were made by the author: a knot. Others were presented ready-made: The *World Book Encyclopedia*. Certain authors reflect on an object's role in a significant life transition—an object serves as a marker of relationship and emotional connection. In other essays, the balance shifts to how an object tied the author to intellectual life—to building theory, discovering science or art, choosing a vocation. In every case, the object brings together intellect and emotion. In every case, the author's focus is not on the object's instrumental power—how fast the train travels or how fast the computer calculates—but on the object as a companion in life experience: how the train connects emotional worlds, how the mental space between computer keyboard and screen creates a sense of erotic possibility.

This collection begins with essays on the theme of discovery and learning, and then, following the arc of the life cycle, the essays discuss the opportunities and challenges of adulthood—the navigation of love and loss—and finally, the confrontation with transcendent issues such as spirituality and the sublime. Life, of course, is not lived in discrete stages, nor are the relationships with objects that accompany its journey. Objects have life roles that are multiple and fluid.

We live our lives in the middle of things. Material culture carries emotions and ideas of startling intensity. Yet only recently have objects begun to receive the attention they deserve.

The acknowledgment of the power of objects has not come easy. Behind the reticence to examine objects as centerpieces of emotional life was perhaps the sense that one was studying materialism, disparaged as excess, or collecting, disparaged as hobbyism, or fetishism, disparaged as perversion. Behind the reticence to examine objects as centerpieces of thought was the value placed, at least within the Western tradition, on formal, propositional ways of knowing. In thinking about science, certainly, abstract reasoning was traditionally recognized as a standard, canonical style; many have taken it to be synonymous with knowledge altogether.

Indeed, so highly valued was canonical abstract thinking, that even when concrete approaches were recognized, they were often relegated to the status of inferior ways of knowing, or as steps on the road to abstract thinking. It is poignant that Claude Lévi-Strauss and the psychologist Jean Piaget, who each in their way contributed to a fundamental revaluation of the concrete in the mid-twentieth century, also undermined the concrete thinking they promoted.[2] Piaget recognized that young children use a style of concrete reasoning that was too efficacious to be simply classified as "wrong." His response was to cast children's "close-to-the-object" approach as a stage in a progression to a formal think-

ing style.[3] Lévi-Strauss recognized the primitive's bricolage as a science of the concrete that had much in common with the practice of modern-day engineers. He said
he preferred to call it "prior" rather than "premature";
yet it was not fully equal.[4]

Beginning in the 1980s, concrete ways of thinking
were increasingly recognized in contexts that were not
easily dismissed as inferior, even and perhaps especially
in the world of science, the very place where the abstract
style had been canonized. Scientific laboratories were
shown to be places where discoveries are made in a concrete, ad hoc fashion, and only later recast into canoni-
cally accepted formalisms; Nobel laureates testified that
they related to their scientific materials in a tactile and
playful manner.[5] To this testimony from science studies
was added the work of feminist scholars who documented the power of concrete, contextual reasoning in a
wide range of domains.[6] Indeed, there has been an increasing commitment to the study of the concrete in a
range of scholarly communities.[7] To this conversation,
Evocative Objects contributes a detailed examination of
particular objects with rich connections to daily life as
well as intellectual practice. Each author has been asked
to choose an object and follow its associations: where
does it take you; what do you feel; what are you able to
understand?

A jeweled pin, simple, European, clearly of the old
country, ties a daughter to her mother and her mixed
feelings about their immigrant status. An immersion in
the comic books of youth teaches a man how to read the
lessons of superheroes in midlife. A lonely graduate student is comforted by her Ford Falcon. The car feels like
her "clothing" in the world of the street, a signal of her
taste and style. When she becomes a mother, it's time
for a trade-in and a BMW station wagon.

Some objects are experienced as part of the self,
and for that have a special status: a young child believes
her stuffed bunny rabbit can read her mind; a diabetic

is at one with his glucometer. Other objects remind us of people we have lost.[8] An artist dies, his collection of Chinese scholars' rocks is left behind. A rock of meditation, "The Honorable Old Man" becomes a presence in the life of his widow, who describes it as she would her artist-husband—"obsession, looking, openness to being surprised and moved, dignity."

Most objects exert their holding power because of the particular moment and circumstance in which they come into the author's life. Some, however, seem intrinsically evocative—for example, those with a quality we might call *uncanny*. Freud said we experience as uncanny those things that are "known of old yet unfamiliar."[9] The uncanny is not what is most frightening and strange. It is what seems close, but "off," distorted enough to be creepy. It marks a complex boundary that both draws us in and repels, as when, in this collection, a museum mummy becomes an author's uncanny "double." Other objects are naturally evocative because they remind us of the blurry childhood line between self and other—think of the stuffed bunny whose owner believes it can read her mind[10]—or because they are associated with times of transition. Transitional times (called "liminal," or threshold, periods by the anthropologist Victor Turner) are rich with creative possibility.[11] In this collection, we follow a young man from the Australian outback as he boards the Melbourne train, finally a passenger on a long-imagined journey. On the train, poised between states of being, everything solid and known can becalled into question

Evocative objects bring philosophy down to earth. When we focus on objects, physicians and philosophers, psychologists and designers, artists and engineers are able to find common ground in everyday experience.

Each narrative in this collection is paired with a short excerpt drawn from philosophy, history, literature, or social theory. The authors of these excerpts range from Lewis Thomas to Umberto Eco, from William James to Susan Sontag. These texts begin to describe the kinds of connections that help us investigate the richness of objects as thought companions, as life companions.

The excerpted theorists engage the essays across a wide range of ideas. I have already noted some. *There is the power of boundary objects and the general principle that objects are active life presences.* Lévi-Strauss speaks of tinkering; Jean Piaget, of the child as scientist. With different metaphors, each describes a dynamic relationship between things and thinking. We tie a knot and find ourselves in partnership with string in our exploration of space. *Objects are able to catalyze self-creation.* When Igor Kopytoff writes about the "biography of things," he deepens our understanding of how a new car becomes a new skin, of how a change of jewelry can become its own voyage to a new world. *Objects bring together thought and feeling.* In particular, objects of science are objects of passion. Essayists who raise this issue are paired with writings from philosophy (Immanuel Kant and Edmund Burke, on nature's sublime) and also from anthropology (Mary Douglas, on the passion behind our need to classify).

I have also touched on the idea that *we often feel at one with our objects.* The diabetic feels at one with his glucometer, as increasingly we feel at one with the glowing screens of our laptops, our iPods, and our BlackBerries. Theorists as diverse as Jean Baudrillard, Jacques Derrida, Donna Haraway, Karl Marx, and D. W. Winnicott invite us to better understand these object intimacies.

Indeed, in the psychoanalytic tradition, both persons and things are tellingly called "objects" and suggest that we deal with their loss in a similar way. For Freud, when we lose a beloved person or object, we begin a process that, if successful, ends in our finding them again, within us. It is, in fact, how we grow and develop as people. *When objects are lost, subjects are found.* Freud's language is poetic: "the shadow of the object fell upon the ego." The psychodynamic tradition—in its narrative of how we make objects part of ourselves—offers a language for interpreting the intensity of our connections to the world of things, and for discovering the similarities in how we relate to the animate and inanimate. In each case, we confront the other and shape the self.

For me, working with these ideas, editing this book, combining the narratives with literary and theoretical texts, and seeing them refracted through different prisms, became its own object discipline, my own practice of bricolage. In this sense, *Evocative Objects: Things We Think With* became for me an evocative object. Its elements were new, but the activity of working on it was familiar, as familiar as carefully handling the objects in the memory closet I knew as a child.

Walt Whitman said: "A child went forth everyday/ and the first object he look'd upon, that object he became." With generosity of intellect and spirit, the authors in this collection engage with the objects of their lives. For every object they have spun a world. They show us what they looked upon and what became the things that mattered.

Objects of Design
and Play

The playing adult steps sideward into another reality; the playing child advances forward to new stages of mastery. I propose the theory that the child's play is the infantile form of the human ability to deal with experience by creating model situations and to master reality by experiment and planning. It is in certain phases of his work that the adult projects past experiences into dimensions which seem manageable. In the laboratory, on the stage, and on the drawing board, he relives the past and thus relives leftover affects; in reconstructing the model situation, he redeems his failures and strengthens his hopes. He anticipates the future from the point of view of a corrected and shared past.

No thinker can do more and no playing child less. As William Blake puts it: "The child's toys and the old man's reasons are the fruits of the two seasons."

—Erik Erikson, *Childhood and Society**

*Source notes for all epigraphs begin on p. 364.

MY CELLO

Tod Machover

My mother tells me that I started music training when I was two. She was my teacher, helping me make music at the piano and find music all over the house. Each week, we set out on expeditions of her devising, discovering household objects that made interesting sounds, that could in turn be combined to create new textures, emotions, and narratives. Then followed the task of making a "picture" of our new composition so that we could recreate it the following week. I learned to invent music from these first principles: sound, structure, score.

As I began to listen to orchestral music (I remember Leonard Bernstein's *Young People's Concerts*), I yearned for an instrument that had the feel of those natural, malleable objects around the house. I wanted my instrument to be able to sing, expressing as much between the notes as on them. The piano, with its special precisions, simply didn't appeal. By the time I was eight, I had chosen the cello, embracing it before learning the details.

Cellos, I found, are the perfect size. Violins are too petite, fingers stepping on fingers; the double bass is a struggle, hands stretched and muscles flexed. But the cello is the size of a human body, reaching the ground as its scroll grazes the top of the head of the seated musician. The cello range is identical to the human voice—that is, the male and female voice combined. The lowest cello note is at the bottom range of a *basso profundo,* and although the cello can actually scream higher than any singer, it has a more normal top range that competes with a diva *coloratura.*

Seated at the cello, my body assumes a calm, natural position—my shoulders relaxed, letting gravity help bow pressure. Yet I can feel the instrument vibrate from head to foot as I draw my bow across its strings, a throbbing through my chest, a buzzing through my legs and feet, a tingling to my fingertips. Sensitive to an extraordinary range of touch, cellos respond to the almost motionless gliding of a gentle *legato* as well as to the he-man crunch of a raspy *sforzato*. The cello is big enough to put up a fight, yet is the largest instrument that you can comfortably carry, or not so comfortably, as I learned when I took it trekking in Nepal and on the New York subway in rush hour.

Unlike the violin that can screech in the hands of beginners, the cello always has a mellow sound and seldom is truly ugly, yet there is an infinite gradation of tone quality and therefore infinite scope for improvement. Because the physical position one takes with the cello is so natural, it is easier to play than the violin and harder than the bass. Both hands and arms are given independence, working in synchrony (something that young players find hard to master) while doing completely different things. The cello is just hard enough, and for me, this gives cellos the right degree of difficulty. And it makes playing cello the perfect companion to thinking. Like walking, playing the cello engages just enough of my mind to suppress internal chatter, leaving me free to imagine.

A similar balance of not too hard/not too easy applies to intonation on the cello, where playing in tune is easier than on the violin (its greater size, quite simply, leaves you more room to find the right note) but still subject to the subtlest inflections. The physicality of the cello is itself slightly irregular, with strings of different thicknesses that vibrate with different degrees of effort, a bridge and fingerboard sloping unevenly under the four strings, and decreased spacing between notes as one goes higher on each string. This means that each note feels different to play. The piano is designed for potential perfection that seems to challenge players to achieve machine-like accuracy. The imprecisions of the Japanese *shakuhachi* are designed so that the player is never certain of exactly how the instrument will respond. The cello stands between these two, pleasurably controllable, yet with pure perfection always slightly out of reach. Very early I realized that lifetimes had been dedicated to exploring and mastering the cello and that one lifetime could never suffice.

In my own case, under my mother's tutelage, I began with the classics and stayed with them—that is, until the appearance of *Sgt. Pepper* when I was thirteen. That album marked my first musical struggle with my mother, who refused to understand how I could like the Beatles. I moved closer to my father, a pioneer in the field of computer graphics and more comfortable with popular culture. I tried to turn my cello into an instrument for composing and performing rock music: I threw away the bow, turned the instrument sideways and propped it on my lap like a (very big, fat) guitar, clamped headphones around its belly, plugged it into a guitar amp and jammed. I tried the same thing with an electric bass guitar, but it lacked the sonic richness, thick-stringed resistance, wide range, and lightning action of my cello. Soon I was improvising and composing, experimenting with tape recorders, multi-track layering, all with this electrified cello.

I managed to cultivate my classical and rock experiences with the cello separately, safely avoiding their collision. That changed when I was sixteen and began to study with a new cello teacher, Richard (Richie) Bock, who played classical, jazz, and rock. Richie destroyed my complacency about music making, beginning with my assumptions about technique. Instead of focusing on the left hand that played notes and mastered intonation, vibrato, and glissandi, Richie put the right hand and the bow in the foreground. The most important, he said, was "the part nobody thinks about, the part that comes easy. The bow is where expression comes from, like breathing for a singer." And furthermore, he said *my* bowing was lousy, so bad in fact that I had to start from scratch.

For months, Richie had me play long-drawn bows over open strings, with no notes played by the left hand at all. I learned to see nuance in cello playing: the constant adjustment of pressure, speed, and angle depending on thickness of string and section of bow; the sweet spot of resonance when the instrument is allowed to vibrate freely; the great beauty that can be found in a simple, constant sound played fully, evenly, purely. By going back to basics, I discovered how to listen carefully and critically, to sense the slightest movement or tension in finger, hand, arm, and back. I learned to meditate in sound. I learned how to practice for real.

By the time I was ready to begin conservatory at Juilliard, I knew I was more interested in composition than performance. Free from thinking of the cello as a profession, I felt I could explore repertoire and my own musical ideas without outside approval. My new teacher, Mosa Havivi, made me rethink what it means to project a musical experience outside of oneself, to hear and feel one's playing as others do. Mosa taught me that I could— and had to—make my own decisions about interpretation and meaning.

As a child musician, the physical intensity of cello playing (a whole body experience, not just a finger activity) had led me to a dissociation of analysis and expression. I performed by ear and feel. Theory was pure abstraction. Now I began to make the conscious connection between thought and touch that had eluded me.

Indeed, there is much in musical education that encourages the dissociation of thought and touch. At Juilliard, Beethoven, a deaf composer, was held up as the ideal composer. Beethoven, the mythology went, was so great that he imagined all his music in his inner ear, not only being unable to hear it in the external world but also shunning the mundane reality of physical vibrations that would dilute the Platonic ideal of his imagined sounds. What this meant at Juilliard was that no composer would be caught dead in a practice room, or plinking out his or her music on a piano, lest he or she be accused of inadequate ear training, of a sterile musical imagination.

But for me this was impossible: my feeling for composition called upon my intimate relationship with the cello. My musical training has separated sound and touch, thought and feeling, concrete and abstract. My relationship with the cello helped me to bring these things together. While at Juilliard I not only sought ways to hear and touch my music as I was composing it but also I began to imagine instruments that could be adapted to the musical requirements of each new project. So I started working with digital computers, learning Fortran (not a popular thing to do at that time, in that environment) in an attempt to model the sounds I was hearing in my head. I also took a four-month trip to India with my cello, traveling extensively, meeting and listening to some remarkable Indian musicians and playing solo Bach suites for them. I began to appreciate the relativity of the cello and of Western classical music; Bach sounded strange to many people I met, and by the time I came home, the cello sounded monochromatic in pitch and timbre to me. I used my new knowledge of computers to produce sounds and textures that went beyond the cello. And I translated my experience with computers and electronics into new playing techniques and compositional experiments for the cello.

After Juilliard, I went to Paris to work at Pierre Boulez's new Institut de Recherche et Coordination Acoustique/Musique (IRCAM). I arrived at a moment when some of the world's first digital synthesizers were being developed. Here, I found my calling—the design of performance and composition systems that could marry the precision of programming with the spontaneity of human gesture. When I came to the MIT Media Lab in 1985, I worked with colleagues to invent instruments (I called them Hyperinstruments) that could enhance virtuoso performance as well as new systems (such as *The Brain Opera*) that could introduce music making to the general public. I designed toys to introduce children to music and thought of my mother and our explorations of sound in our home. In the mix of new instruments and musical forms—rhythmic Beatbugs, squeezy Music Shapers, and the sinuous Melody Easel—my inspiration has always remained the cello.

Coming full circle to music and childhood brings me to my own two daughters—Hana, now 12, and Noa, now 8. They are studying music—and although they do like playing Beatbugs and composing with Hyperscore, musical technologies of my invention, Hana is learning violin and Noa piano. I practice with each of them every day, trying to keep what was good about my Mom's coaching. The violin is just different enough from the cello that it keeps me on my toes. How do I teach a slide, a note perfectly in tune, a bow beautifully changed, a phrase delicately shaped, a musical story deeply felt and meaningfully conveyed? How do I share my love of music with my daughters when there is so much tough technique to learn, so much frustration to overcome? How do I reconcile the desire to build computational music toys that convey immediately the excitement and joy of music making with the need for practice and discipline and experience that can only mature over a lifetime?

My daughters' fits and starts with music have helped me to return to the cello with a fresh perspective. These days I do not perform on it often, but I do use the cello to try out new ideas. When a period of musical work is ending and I feel a new one beginning, I like to let my ideas percolate in my imagination, but I also like to touch them, and the cello is my tool for that. I try out new sounds that stimulate my physical memory: when

I hear melodies or intervals, I can feel what my left hand fingers would do to create them. When I am imagining—in the quiet of my study—a full orchestral sonority, my muscles reproduce the gesture as if I were playing it on the cello.

And still, perhaps above all, I play the cello to concentrate, to meditate, to relax. It remains for me the perfect gauge of complexity, of how much an individual human being can shape or master, follow or comprehend. Playing the cello remains the activity that I do best and that I do only for myself. It is the object that is closest to me that I don't share with others, the intermediary I use to reconnect to the forces and feelings that drew me to music in the first place.

Tod Machover, composer, inventor and cellist, is Professor of Music and Media at the MIT Media Lab.

(very gentle)
[1'08"]
sempre p
Gliss.
possible on C-string
FADE IN "Harmonics";
FADE OUT "Inharmonic Drones".
etc.
[1'24"]
Harmonic Bounce
Harm. Bounce
as high as possible on G-string
mp
[1'35"]
Play fast and fluttering material using harmonics in high register on D-string; move in and out of "noise," in and out of tremolo—imitate computer part
legato
rhythmic but delicate
[1'45"]
[1'47"]
[1'51"]
[1'53"]
slap bounce
(no break)
gentle but irregular glissando
[2'03"]
dolce
low, indistinct "thrash"
as high as possible on A-string; gentle noise tremolo
To absolute stillness...
[2'13"]
(noise)
Fade out with computer part

The "bricoleur"['s]. . . universe of instruments is closed and the rules of his game are always to make do with "whatever is at hand." . . . Further, the "bricoleur" also, and indeed principally, derives his poetry from the fact that he does not confine himself to accomplishment and execution: he "speaks" not only *with* things, as we have already seen, but also through the medium of things: giving an account of his personality and life by the choices he makes between the limited possibilities. The "bricoleur" may not ever complete his purpose but he always puts something of himself into it.

—Claude Lévi-Strauss, *The Savage Mind*

KNOTS

Carol Strohecker

I remember the day I taught my younger brother how to tie his shoes. I was nine years old and he was three, and since I often looked after him, I also frequently found myself tying his shoes. That day, we sat together on our staircase, our legs bent toward us. Looking down at our shoes, I remembered how a little mantra had helped me learn to write a figure 5: the pencil went "down, around, hat" and in three strokes reliably produced the numeral. So I made up a mantra about shoelaces having something to do with left, right, loops, and around, which I recited while moving the pieces of string accordingly—first on my foot and then on his.

My brother's excitement grew as he observed me and then tried the technique for himself, repeating it until it worked and resulted in a triumphantly tied pair of shoes. His excitement reflected my own as I marveled not only at his diligence, but at the power of the simple mantra. Watching him carefully looping his laces, I saw myself mirrored in a younger child.

My own knot work developed through my teens as I generated macramé designs for belts, bracelets, potted plant hangers, shawls, room decorations, and the like. I adorned my siblings and friends, my walls and keepsakes with knots—in chains, braids and spirals, and with all manner of string weights, textures, and colors. I calculated lengths and costs; mastered arm bends, wrist flicks, hand spans, and fingertip maneuvers; and learned to see things dimensionally, imagining repetitions, alternations, interspersals, and entwinements. I didn't know I was beginning to think like a mathematician. I was simply having fun. I enjoyed generating the creations and seeing how people received them.

"Knot Lady" was a name I first earned from the children I worked with at the MIT Media Lab. After entering graduate school at MIT, I created a Knot Laboratory where I taught children, most of them around ten years old, to tie knots and talk to me about their experiences. Over a year, we transformed a bleak, urban classroom into a lively laboratory space devoted to learning with knots.[1]

Each day at school, I was greeted with a large sign: "KNOT LAB." Constructed by three students who mixed string knot formations with pictures of a chemist's flask and party balloons for its design, the sign reminded me of the simultaneously playful and serious business that took place behind those doors.

Inside our "Knot Lab," children played with string, tacked knots onto display boards, and worked together on stories about knots. The products of their experiments— large, colorful displays of knots in various stages of formation were drawn on paper, tacked to walls, and dangled from the ceiling.

Dozens of knot forms found their home in the Knot Lab. They included simple knots like the Overhand, Figure 8, and Stopper; square knots like the Stevedore and Granny and Thief; and movable knots like the Running Bowline, True Lovers', and Trumpet. To construct them, the children considered unknots, tangles, mirror images, handedness, and knotty spatial relations—over, under, around, and between. They wrapped, rotated, flipped, twisted, and shifted scales as they tied. Their thinking spanned the deliberate and spontaneous, the rational and affective, the conscious and unconscious. And individual preferences were apparent: some children dealt with a knot as an integral entity produced by moving a single end of the string; others broke the process into steps, following and creating procedural instructions; and still others combined pieces—smaller knots as modules—to build up more complicated knots. These approaches were each productive, but they were also very different. The knots demonstrate the diversity (rather than the standardization) of styles of learning. They are objects that enable us to explore the inner states of those who tie them.

One of the most avid knot-tyers was a girl named Jill. I remember that she tended to be serious in the lab, that she was neat and polite, and that she liked to sit close, touch, and talk at length about the knots she worked on. She liked being reassured about her work, which was careful and deliberate. What she didn't like was to leave something unfinished. She stayed with her projects until they were done and tried to convince others to do the same. She didn't like to skip steps; she wanted the sense of accuracy that only the careful progression from one detail to the next could provide.

I noticed early on that more than for any of the other lab participants, it was important to Jill to designate clear anchor points for the string as she tied new knots. On the way to producing a knot, she would often resort to stapling or taping down parts of the string. It was important to Jill to articulate and anchor intermediary configurations, in order both to understand a knot and create a record for later reference.

As the project progressed, Jill told me that her parents had recently divorced, and that she and her brother lived half of the week with their mother and half of the week with their father. She mentioned that there was tension in her parents' communication and that it troubled her. She told stories of situations in which any reasonable action on her part would have slighted one of her parents. She seemed to feel herself in a perpetual "double-bind," doomed to doing something wrong no matter what she chose, torn between decisions that her parents might see as representing the interests of one or the other of them.

Jill was absorbed with knots whose completed state involved motion. She once spent days creating an exhibit of such knots, where passers-by could pull the ends of a True Lovers' knot she had suspended from a pipe on the ceiling in order to play with the knot's back-and-forth movement. Jill made several iterations of the knot before the exhibit took its final form, modifying the string to facilitate pulling its ends. To hang her construction, she anchored a long string to a ceiling pipe with a Square knot. A Bowline at the end of this string held one of the two strings composing the True Lovers' knot, which supported the second string wrapped around it. Excited about her construction, she made a "museum label" highlighting the placement of the three knots:

> At the very top [on the black pipe] notice the "Square knot" to hold it in place. The knot holding on to the Lovers' Knot [True Lovers' knot] is the "Bowline." Notice the way the strings are two colors. It is that way so it is easier for you to pull it.
>
> To pull take the two strings with the black Lego pieces. Pull hard until the two pretzel knots meet. Then pull hard the two strings without anything on them. Repeat if you wish.
>
> Please pull me.

To me, Jill's final phrase signaled her identification with the knot. And it seemed to echo another voice in her mind that wanted to say: Notice how I am suspended by two knots, one that anchors me and one that holds me. Notice how I am two knots, waiting to be pulled this way and that. I understand being pulled; it is something that I know. Allowing others to pull me is a purpose that I serve.

Through the course of the project, Jill expressed her emotions in knots and tried to initiate some emotional repairs as well: frustrated with being pulled by others led her to devise a step-by-step approach to knot tying. Others might leave; Jill committed herself in advance to a plan.

Six years after the Knot Lab had closed, I was able to find Jill and another member of the original project. They were curious about reconnecting with each other and with me. Jill remembered me as the "Knot Lady" but claimed not to remember much about knots. I thought that in this she was expressing her anxiety about mathematics. Although Jill had been one of the most avid participants in the Knot Lab when she was younger, in the intervening years she had come to think of herself as a person who was "not good at math," a self-image all too common among young women. Jill was open to discussing her lab experience and to participating in new projects involving colorful polyhedra but hesitated when our explorations involved some numeric quantification of an idea. The gap between what she could do and what she thought she could do was poignant.

It may be that I am the one for whom the Knot Lab had the most impact. Knot making showed me how commonplace objects can help people think purposefully about continuity and separation, combination and deviation. Through knots I learned that engaging objects can help people to build intuitions about mathematics. And witnessing one of the female participants succumb to stereotypical math phobia after such a strong start as a fifth grader spurred my determination to encourage the representation of different learning styles in all pedagogy.

For many, however, I will always be simply the Knot Lady. My growing collection of knot-oriented gifts serves as constant reminder of this: a ceramic vase with a Square knot decoration and braided handles, a clock with knots in places of numbers, two seared glass spindles entwined to form an elegant bracelet. And new objects and e-mails continue to come my way from people whenever they encounter news about knots—whether it's an article about the usefulness of knot theory in

DNA research, a publication from *The Shipping News,* or endearing knot jokes. In truth, I wouldn't want it any other way. Much as painters relish a blank canvas, writers a fresh page, or moviemakers a darkened screen, I suppose I will always have a penchant for bits of string and the potentials they suggest.

Recently, I asked my brother if he had any memories of learning to tie his shoes. He told me he recalled a moment when he had just completed tying his shoes and left the house to join his friends. I like to imagine that this moment occurred after he mastered the strings and mantra on the stairs, only steps from the front door of our house.

Carol Strohecker was Principal Investigator of the Everday Learning Research Group at Media Lab Europe, and is now director of the Center of Design Innovation, an institutional partnership of the University of North Carolina.

[Electronic communication] . . . is on the way to transforming the entire public and private space of humanity, and first of all the limit between the private, the secret (private or public), and the public or the phenomenal. It is not only a technique, in the ordinary and limited sense of the term: at an unprecedented rhythm, in a quasi-instantaneous fashion, this instrumental possibility of production, of printing, of conservation, and of destruction of the archive must inevitably be accompanied by juridical and thus political transformations. . . . [Because of] these radical and interminable turbulances, we must take stock today of the [archived] classical works. . . . [C]lassical and extraordinary works move away from us at great speed, in a continually accelerated fashion. They burrow into the past at a distance more and more comparable to that which separates us from archaeological digs.

—Jacques Derrida, *Archive Fever*

THE ARCHIVE

Susan Yee

La Fondation Le Corbusier in Paris archives the work of the world-renowned architect, Le Corbusier. His work is studied by every student of architecture, and in the mid-1990s my task was to closely examine his sketches, drawings, notebooks, models, anything I could find that might help to construct a virtual model of one of his famed unbuilt projects, the Palace of the Soviets. The archives were located in Le Corbusier–designed buildings, Villa La Roche and Villa Jeanneret; the idea of sifting though the master architect's original drawings in a space that was conceived by the master himself thrilled me. The materials were rich: fluid sketches, detailed drawings, study models, and notes. I read his letters. I browsed through his datebook and imagined his days full of meetings. I examined his hand-scrawled calculations in the margins of sketches and did the math along with him. There were newspaper clippings. I remember finding one where his design was critiqued. Right on the clipping he had written "Idiote" in a vigorous and powerful hand. I could trace the precision and force of the incision into the newsprint. I felt his frustration, his spirit.

One day, I asked to see the overall plan drawing for his unbuilt design. I was escorted to a special room where Le Corbusier's largest drawings were viewed and waited for the curator to bring up the large rolled drawing. I waited in silence as the curator opened the scroll. It was so large that it spilled over the edge of the table. I had to walk around the drawing in order to see it. I expected to be given gloves, but I was not. I felt awkward. I stood there more than timid, almost paralyzed. I didn't know if I could or should touch it. And then the curator touched it, so I went ahead and touched it too with my bare hands. All I could think about was that this was Le Corbusier's original drawing. It was meticulously hand-drawn, but the drawing was dirty. There were marks on it, smudges, fingerprints, the marks of other hands, and now I added mine. I felt close to Le Corbusier as I walked around and around the drawing, looking at the parts that I wanted to replicate to bring home with me, touching the drawing as I walked. The paper was very thin.

The next day I came back to the archive and that same scroll was rolled out again. The ritual began again. I spent all day walking, touching, looking, thinking. On other days the ritual would be different. I looked at Le Corbusier's personal, handwritten letters. And one day, and this was the most miraculous of all, I found a little parchment bag full of paper squares of different colors and different sizes. I was there with a team of other MIT architects, and we all gravitated toward these playful cut-outs. Delighted with the discovery, we all immediately came to the same idea at once: that these were the elements Le Corbusier used when he was designing the Palace of the Soviets. These were the little squares he used to program the large project. He figured out the arrangement with little colored papers. One color was for meeting rooms, another was for public areas. Each function of the project had a designated color. And I imagined how he fiddled with these little bits of paper until he found a programmatic configuration that pleased him; I fiddled with them too.

On my last day at the archives, the curator approached me with pride, "Oh, you'll love what we're doing now. You won't ever have to come here! You won't ever have to look at these drawings anymore! We're putting them all in a digital database!" She brought me to an adjacent room and showed me the exact drawing I had been looking at, the drawing around which I had been circling for days. It appeared on her computer as a small icon. If you clicked on it, it became larger. If I had accessed this drawing from home, I would never have grasped its dimensions, I would never have known that it was stored separately, carefully rolled, that it was dirty with smudges and fingerprints. The scans for the Web site gave me nothing to touch. I felt no awe about the scale of the drawings. Looking at the curator's scans made me think respectfully about mass consumption, about allowing everybody to have access, about the technical problems of how to use a cursor to move around the drawing on the screen, and about how differently I understood the digital image and the designer behind it.

Looking at the scans in the computer room made me miss the quiet of the physical archive, the ritual of bringing out the precious original drawings, the long minutes of unwinding. Sitting at the curator's computer in Paris, I followed her instructions and linked once again to the drawing. A moment later, some bit of business crossed my mind and I linked to MIT. Feeling like a saddened citizen of the information world, I felt transported to MIT through the link. I had a moment of shame.

That day with the curator was the first time I began to think about the transition from physical to digital. The evocative object, the Le Corbusier drawing in both its physical and digital form, made me wonder how automatic it had been for the curator to put the emotion of the archive out of mind, how easy it was to trade the value of touch and physicality for the powers of digitization.

I think of Turkle's distinction between instrumental and subjective technology, between what technology does for us and what it does to us as people.[1] The new Le Corbusier digital database did things for me. It allowed me to do things that I could not do before. I could search it, manipulate it, copy it, save it, share it. But what did it do to me? It made the drawings feel anonymous and it made me feel anonymous. I felt no connection to the digital drawings on the screen, no sense of the architect who drew it.

As I came to terms with my anonymity, my lack of connection, and the loss of my former rituals in the physical archive, I felt fortunate to be in a generation of designers that straddles both physical and digital worlds, a generation that creates, values, and understands handmade drawings and models as well as digital ones.

In my work designing technology-enhanced studios at MIT, I often think about Le Corbusier's drawings and the drawings that designers make today. Today's drawings and models are constructed on the computer. They have never been physical. They are born digital. They will never be touched. I think about how a new generation will be trained to favor computational techniques and algorithmic methods of design. Instrumentally, these technologies offer opportunities for innovation in design development and construction. Subjectively, however, what will these technologies do to us? How will they affect the way we feel, see ourselves, and see design? How will future students of architecture come to experience the designs of a master from the pre-digital era? And what of the "old masters" of our first digital era? Will future students be satisfied to simply understand the algorithms that generated their designs? Will we still crave some pilgrimage such as the one I took to Paris? But there will be no place to go; it will all be on a collection of servers. What will this do to our emotional understanding of the human process of design? What rituals might we invent to recover the body's intimate involvement with these new traces of human imagination? Will we be able to feel the human connection through digital archives? Will we care?

Susan Yee earned a PhD in architecture from MIT and studies the implications of integrating new technologies into design learning environments.

To express the same idea in still another way, I think that human knowledge is essentially active. To know is to assimilate reality into systems of transformations. To know is to transform reality in order to understand how a certain state is brought about. By virtue of this point of view, I find myself opposed to the view of knowledge as a copy, a passive copy of reality. In point of fact, this notion is based on a vicious circle: in order to make a copy we have to know the model that we are copying, but according to this theory of knowledge the only way to know the model is by copying it, until we are caught in a circle, unable ever to know whether our copy of the model is like the model or not. To my way of thinking, knowing an object does not mean copying it—it means acting upon it. It means constructing systems of transformation that can be carried out on or with this object. Knowing reality means constructing systems of transformations that correspond, more or less adequately, to reality. . . . Knowledge, then, is a system of transformations that become progressively adequate. . . . But let us ask what logical and mathematical knowledge is abstracted from. There are two possibilities. The first is that, when we act upon an object, our knowledge is derived from the object itself. . . . But there is a second possibility: when we are acting upon an object, we can also take into account the action itself, or operation if you will, since the transformation can be carried out mentally. In this hypothesis the abstraction is drawn not from the object that is acted upon, but from the action itself.

—Jean Piaget. *Genetic Epistemology*

STARS

Mitchel Resnick

When I was growing up in a suburb of Philadelphia, there was a small field on the side of our house.[1] On summer evenings, I would go to the "side lot" (as we called it), lie on my back, and stare into the sky. My eyes would dance from star to star. But it wasn't so much the stars that held my attention. Rather, it was the space between, around, and beyond them. At an early age (maybe seven or eight), I had started to wonder about all that space. Does it go on forever? If not, where does it end? How does it end?

Every answer that I could think of seemed equally absurd. I could not imagine the universe going on forever. But how could it end? If there is a wall at the end of the universe, what is on the other side? These questions frustrated and fascinated me. Of course, I came across many other questions that I couldn't answer. But for most questions, even if I didn't know the answer, I could at least imagine that there *was* an answer. Questions about the "end of the universe" took on a special status for me. These were questions where I couldn't even imagine any answer. No answer seemed possible.

As I grew older, I became interested in puzzles and paradoxes. I spent many hours trying to sort out the sentence: *This sentence is false.* If the sentence is true, then it must be false. But if it is false, it must be true. Again, a puzzle for which I couldn't even imagine any answers.

In school, I was attracted to math and physics, two fields filled with paradoxes and counterintuitive ideas. I became fascinated by an object that my high-school physics teacher showed us. The object was remarkably simple: two wheels and an axle, with a pin hanging down from the middle of the axle (not quite hitting the ground), and a string at the end of the pin. The teacher asked: What happens when you pull on the string? Since the string is attached to the end of the pin, it seems that the pin should come toward you. At the same time, it seems that the wheels should come toward you. Both can't be true: if the pin comes toward you, the wheels move away; if the wheels come toward you, the pin moves away. Another paradox! But this object was different from the stars of my childhood: you could hold it in your hands and test it out. Indeed, I went home, took apart an old toy truck, and made my own version of the puzzle, testing pins of different lengths. Even after I "knew" the answer, I loved tugging on the string and thinking about the paradox.

In college, majoring in physics, I was determined to develop a better understanding of what I now thought of as my Ultimate Paradox—the paradox of a universe that can't go on forever but can never end. In physics courses, I learned how to derive and manipulate the equations of general relativity, the field most directly related to my paradox. It wasn't the equations that really interested me, they were just a foundation, a jumping-off point, for thinking about the paradox itself. I tried to approach it through new thinking strategies, through new intuitions and metaphors: I learned that the universe might curve back on itself, just as the land on Earth curves back on itself as you travel all the way around the globe. But what does that mean? How can three-dimensional space "curve back on itself"? How could I envision that? How could I "feel" that?

During college, I had planned to attend graduate school in physics. But at the end of senior year, I decided to work as a journalist instead. I worried that physics graduate school would be filled with too many equations and too few qualitative insights. I was still fascinated with the mysteries and paradoxes of science. I hoped that as a journalist, specializing in science and technology, I would be able to share my fascination with others. For five years, I covered universities and high technology companies around Boston and then Silicon Valley. I enjoyed my work, but something was missing. I didn't feel the same level of intellectual excitement that I had felt in college. I had lost contact with my obsession. I began to recognize the importance of having obsessions.

Then, in 1982, I wrote a cover story for *Business Week* magazine about research in the field of artificial intelligence. I talked with many leading researchers in the field. I became increasingly interested in questions about the mind. How can a mind emerge from a collection of mindless parts? It seems clear that no one part is "in charge" of the mind (or else it too would be a mind). But how can a mind function so effectively and creatively without anyone (or anything) in charge?

At last, I had a new Ultimate Paradox, a new obsession. I wasn't so much interested in the details of neuroscience, or even in traditional research in artificial intelligence. Rather, I wanted to develop qualitative ways to think about the idea of emergence. I became interested not only in minds but also in other systems in which complex patterns emerge from simple interactions among simple parts. I became particularly interested in natural selection and evolution, hoping to gain a better understanding of how today's sophisticated life forms evolved from a few simple chemicals. For me, there was something intriguing and beautiful about this self-organized emergence of order from disorder, of complexity from simplicity. I developed an emotional investment in this idea. Few things got me more upset than listening to creationists attacking the idea of evolution, attacking the idea that complexity can arise from simple pieces.

Around this time, I came to MIT for a year as a Knight Science Journalism Fellow. During the year, I studied with Sherry Turkle, who studied the emotional power of things we think with, and Seymour Papert, who described how a particular object, gears, had changed his way of thinking in childhood. Papert had fallen in love with gears and, in the process, with mathematics.[2] Most important during that year was the way I came to see the computer in a new light. For me, the key insight was not that the computer itself is an evocative object (although surely it is for many people), but rather that the computer can be used to *create* evocative objects. And those new evocative objects could be used to help people learn new things in new ways. In designing the Logo turtle, for example, Papert had explicitly attempted to make an evocative object to help students become engaged with mathematical ideas and mathematical thinking. Just as the young Papert had fallen in love with mathematics through gears, children could now fall in love with mathematics through the turtle.

The idea of creating evocative objects for educational purposes is not a new idea. When Friedrich Froebel started the world's first kindergarten in 1837, he carefully designed a set of physical objects—blocks, balls, beads—that became known as Froebel's gifts.[3] As children playfully experimented with Froebel's gifts, they learned important ideas about number, shape, size, color. This approach has stood the test of time, and it continues as the basis for kindergartens around the world today.

The computer provides an opportunity to expand Froebel's approach, making possible a wider and more diverse range of evocative objects for education. I felt a new sense of mission: I could use the computer to create evocative objects for exploring my new Ultimate Paradox, the paradox of a complex whole arising from simple parts. I wanted to create objects that would enable me to explore the paradox, but also to help others explore it as well. I decided to use Papert's turtle as the basic building block. But instead of a single turtle, I created thousands of turtles. And I developed a new language, called StarLogo, that enabled students to program each of the individual turtles, then observe the patterns that emerge from all of the interactions.

Students have used StarLogo to explore a diverse range of phenomena. They have turned turtles into birds to explore how flocking patterns arise; into cars to explore how traffic jams form; into ants to explore how foraging patterns emerge; and into buyers and sellers in a marketplace to explore how economic patterns form. It has given me great satisfaction to see students become engaged with my Ultimate Paradox. For some, it has become an obsession, as it was for me.

Over the past twenty years, my research has continued to revolve around the creation of evocative objects for education. Working with the LEGO Company, I've embedded electronics inside LEGO bricks, so that children can make their LEGO constructions come alive—sensing, reacting, and even dancing with one another. I aspire for these "programmable bricks" to serve as a Froebel gift for the twenty-first century. Just as the stars of the night sky inspired, intrigued, and provoked me as a child, my hope is to create new objects that help others find their own obsessions.

Mitchel Resnick is LEGO Papert Professor of Learning Research and Director of the Lifelong Kindergarten research group at the MIT Media Lab.

When the stick (hobbyhorse) becomes the pivot for detaching the meaning of "horse" from a real horse, the child makes one object influence another semantically. He cannot detach meaning from an object, or a word from an object, except by finding a pivot in something else. Transfer of meanings is facilitated by the fact that the child accepts a word as the property of a thing: he sees not the word but the thing it designates. For a child the word "horse" applied to the stick means "there is a horse" because mentally he sees the object standing behind the word. A vital transitional stage toward operating with meaning occurs when a child first acts with meanings as with objects (as when he acts with the stick as though it were a horse). Later he carries out these acts consciously.... In play a child spontaneously makes use of his ability to separate meaning from an object without knowing he is doing it, just as he does not know he is speaking in prose but talks without paying attention to the words. Then through play the child achieves a functional definition of concepts or objects, and words become parts of a thing.

—Lev Vygotsky, *Mind in Society*

KEYBOARDS

Howard Gardner

On July 11, 2003, I turned sixty. In front of the twenty or so friends and family that were gathered, my four children gave presentations—a poem, a newly composed piece performed on the piano, and a set of written reflections. I was touched, grateful, and struck by the fact that all four of my children spoke about keyboards. Two described the importance of music and the piano in their (and my) life; two evoked the experience of listening to me type manuscripts at night as they were nodding off to sleep.

I am not sentimental about objects. I admire beautiful things and like to be around them, but I make no effort to purchase or keep them. I am happy wearing clothes of forty years ago; truth to tell, I am happier wearing such old clothing. I save my feelings for other human beings and the family dog, Nero. What I do value are the sounds of music and the ideas in books. If I could no longer hear music (or play it), I would be devastated. If I could no longer read for study or pleasure (or write for others or myself), I would not enjoy life. My preferred access to linguistic and musical objects is via fingers on keyboards.

I began both piano lessons and the typing of manuscripts when I turned seven. I took piano lessons for almost six years, then began to teach piano sporadically, and later sat in on lessons and practiced with two of my children. In all, I have played piano off and on for fifty-three years. While I never learned to touch type, I have seen myself as a writer since second or third grade, and typewriters have been with me ever since. I began with manual typewriters for home and office; over time, I moved in turn to electric typewriters, office PCs, and a succession of laptop computers, on one of which I am typing at this moment.

There are scarcely any days on which I do not move my fingers across some kind of keyboard, and often I am at a keyboard, working on music or writing, for as many hours as I eat or sleep. I am able to write with a pen or pencil, and sometimes do so, but I much prefer to type.

Some people are students of keyboards; they sample hundreds of pianos and prefer only a certain model, one Steinway above all the rest. Others love the keyboard on their computer because of its touch—they feel that their hands glide over it with no wasted motion. But for me, this is not the case. I pay essentially no attention to the quality of my keyboards. All of my attention is focused on the message, musical or literary. When I play the piano, I try to use an appropriate touch, but I am really studying the music, trying to understand it, hoping to capture that meaning through my meager technique. When I am typing, my mind is entirely on the contents that I am trying to convey. Above all else, I am trying to be clear; secondarily, I am trying to write in a way that is pleasing to read.

And yet, even with my focus so intently on the message, the experience of my fingers on keyboards feels like more than simply a means to a desired end. In the creation of both music and text, if I could bypass the keyboard and directly transmit mental signals to an instrument or to the computer, I would not want to do so.

When I learned to play the piano, my mother sat next to me nearly every day. When two of my sons began to play, I naturally sat next to them as well. I feel an association between the piano keyboard and family love. In the case of writing, the sensations of fingers on keys are soothing in a way that goes beyond my pleasure in what I write: when I want to imagine myself happy, I think of myself in my study or in a comfortable hotel room on the road, or even cramped, as I am now, in the economy class of American Airlines Flight 1367, from Boston to Miami, fingers on a keyboard, letting my thoughts proceed at their pace into a typescript.

Stepping back from my personal experience, I don for a moment the perspective of the social scientist. As a scholar, my task is to master the knowledge of the past and to identify ways in which I can add to it. As a sometime pianist, my task is to understand the explicit and implicit instructions of the composer and, ultimately, to introduce my personal interpretation of his or her composition.

In principle, both of these assignments can be tackled simply by thinking about the challenge at hand and arriving at the best possible solution. Indeed, Mozart is said to have created entire compositions in his head and simply to have written them down in the manner of an amanuensis; and various writers have claimed that the job of writing is simply the transposition to paper of words and ideas that have come to them in a flash.

I am skeptical of these accounts. Research on creativity reveals that, even though new ideas appear to come to one as a flash, there has invariably been tremendous preparation beforehand—and this preparation can be documented in the written record. Moreover, thought does not take place in a vacuum—it takes place in various media of expression. By the time one has become an expert some of these media appear to be largely cerebral. But especially during early development, as the social psychologist Lev Vygotsky has taught us, these media are invariably tools that have been created by the wider society—tools ranging from words to pencils, computers, and musical instruments.

Perhaps as an expert writer and a long time journeyman pianist, I could achieve some of my goals without a keyboard. However, I think that I would be handicapped by the absence of an instrument on which to work. And I know that affectively I need, enjoy, even love the opportunity to type or play away, for much of the day.

Howard Gardner is the John H. and Elisabeth A. Hobbs Professor of Cognition and Education at the Harvard Graduate School of Education.

Objects of Discipline
and Desire

There was a little girl who was so delicate and charming, but in the summer she always had to go barefoot because she was so poor. . . . The little girl's name was Karen. . . .

Karen was . . . to have new shoes. The rich shoemaker in town measured her little foot. . . . In the midst of all the shoes stood a pair of red ones just like the ones the princess had worn. How beautiful they were! . . . [Karen] . . . put them on. . . . And Karen couldn't help herself, she had to take a few dance steps. As soon as she started, her feet kept on dancing. It was as if the shoes had taken control. She danced around the corner of the church, she couldn't stop herself. . . . At home the shoes were put in a cupboard, but Karen couldn't help looking at them. . . . She put on the red shoes. Why shouldn't she? And then she went to the ball and began to dance.

But when she wanted to turn right, the shoes danced to the left, and when she wanted to move up the floor, the shoes danced down the floor, down the stairs, along the street, and out the town gate. Dance she did, and dance she must, right out into the dark forest.

—Hans Christian Andersen, "The Red Shoes"

BALLET SLIPPERS

Eden Medina

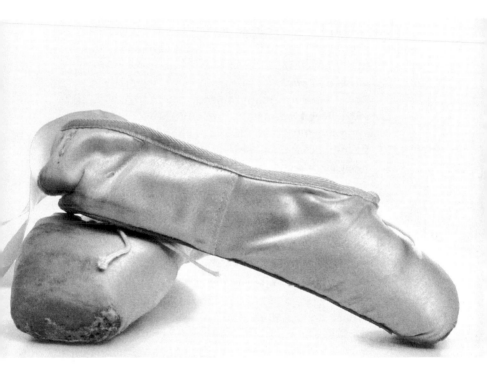

As a child, I lived to dance. My early ballet lessons still stay with me, a long series of carpools from one musty studio to another. I began my training at age four after my parents presented me with my first pair of ballet slippers and drove me to the local studio. Dressed in baggy leotards and pink cotton tights, my fellow four-year-olds and I learned to rotate our hips unnaturally outward into "first position," stand rigidly with our shoulders back and our stomachs sucked in, and eventually associate meaning with French words such as *tendu* and *plié*.

There are many objects associated with ballet, most of which contribute to a culture of continuous self-appraisal (the barre, the elastic band around the waist, the mirrored room). Among these, the shoe is by far the most significant. It acts as an object of identification, drawing a line between the various styles of dance. To a surprising degree, its constraints and affordances define the movement of the ballerina.

History illustrates how the evolution of ballet paralleled the development of the dancer's shoes. Prior to the eighteenth century, this fledgling art celebrated male athleticism and relegated female dancers, clad in heavy skirts, wigs, and heeled slippers, to peripheral roles. French dancer Marie Ann Cupis de Camargo was one of the first women to cross the gender barrier when she removed the heels from her slippers and began performing the same flashy steps as her male counterparts. In 1832, dancer Marie Taglioni forever altered ballet technique by dancing *en pointe* the full-length ballet *La Sylphide*. The resulting performance—ethereal and light—embodied the spirit of the Romantic Age. Women who seemed to possess supernatural beauty and purity captured the hearts of ordinary, earthbound men. Taglioni's portrayal of a weightless, idealized femininity made her an international favorite. It was reported that some overly zealous fans ate her discarded slippers with sauce. In the nineteenth century, choreographers continued to showcase the technique of the female ballerina, who had since displaced the male dancer as the central figure in ballet. The invention of harder, more durable slippers increased the ballerina's potential for athleticism and broadened the range of movement she was expected to perform.

Beyond these technical and aesthetic expectations, ballet shoes carry symbolic power. In the early twentieth century, Isadora Duncan rejected the rigidity of nineteenth-century ballet by donning loose Grecian robes instead of corsets and embracing the naturalness of the bare foot instead of the artificiality of the ballet slipper. Modern dance pioneer Doris Humphrey later based her style of fall and recovery on the movement of the unsoled human footfall.

By the age of eleven, before I had even reached the age of going *en pointe,* I had already disfigured my feet. The restrictive nature of the shoe, combined with the demanding movement required of my feet within them, resulted in numerous trips to the doctor for ingrown toenails as well as the initial signs of bunions. My legs developed the hyper musculature characteristic of dancers forced to raise their bodies up on their toes. I still bear the marks of my early years in ballet.

Yet such inconvenience seemed minor in comparison to my dream that my body might recreate the movements of controlled beauty characteristic of the dance. My ballet slippers enabled me to move in ways I never dreamed possible. I could mimic the ethereal weightlessness of *Giselle* or throw myself into a series of athletic jumps and turns that left me happily gasping for air. For a time, I felt my body would respond to any demand I could impose on it.

In ballet, shoes shape physical artistry and also mark the dancer's progression within the ranks of the discipline. When I was four, my parents purchased my ballet slippers in a mall. They were inexpensive, cut from coarse leather, and were sold with the elastic strap fully attached to the sides of the shoe. My next pair came from a store specializing in dance apparel. Apart from their origins, the most noticeable difference was the piece of unattached elastic my mom had to sew on the slipper, specially positioned to accommodate the dimensions of my foot within this particular shoe. As I improved, I became more demanding of my increasingly sophisticated equipment. I remember my pride when I could finally attach a pair of ribbons to my slippers in addition to the requisite elastic.

A dancer receives her first pair of *pointe* shoes, toe shoes, at age twelve, roughly corresponding to the age she enters puberty. Progress continues to be marked though a progression of shoes, now all shoes for dancing *en pointe*. The new hierarchy is even more complex, marked by technical terms such as "shank stiffness" and "box size." These new shoes also come with an array of accessories, such as first-aide tape and lamb's wool to ease the inevitable pain of blisters, bunions, and bleeding.

My own journey through the hierarchy of shoes signified an increase in my skill and helped me identify with the image of the professional ballerina that I upheld as my physical ideal. For a time, both my movements and appearance progressed along what I imagined to be a natural trajectory toward this goal. However, as I continued my studies, there was an increasing gap between the reality of my body and the perfected body imagined in my mind. My shoes endowed my body with the theoretical capability to balance and extend my limbs, but my legs were not as long, my torso not as limber, and my neck not as graceful as the one owned by my imagined self, my rival.

Eventually each movement I executed before the mirror forced me to stare at my own limitations. Just as the ballet slippers of my youth helped me become a member of a community driven to transform the body into art, the toe shoes of my young adulthood highlighted both my technical progression and the elusive nature of my ideal physique. As I became closer to my ideal in the realm of technical movement, I was left with a profound sense of my physical shortcomings. My body would never be beautiful in the exact way I longed for it to be.

I quit ballet shortly after this realization. I felt that my body had failed me. I put my toe shoes in a box and there they collected dust for the next ten years. As I entered adulthood, the library replaced the ballet studio as my favorite haunt; the computer became my preferred tool of self-expression; and the academic community offered a new mirror for self-appraisal.

Despite my prolonged absence from the dance studio, the movements of my youth remained engraved in my body. Several years ago, I felt an urge to revisit them. My father was able to locate my old pair of ballet slippers, which he promptly shipped to me via FedEx. As I sat in the studio on my first day of class and began to put on my warm-up clothes, I doubted my decision to return: How would the mirror evaluate my older, less flexible body? Yet, as I looked around the studio, I noticed that none of my classmates resembled the ideal that had driven me from the discipline I once loved. Slowly I slipped my feet into my shoes and began to stretch, feeling my hips rotate almost imperceptibly outward as they recalled a stance once second nature. I sensed that whatever the shortcomings of the body, I was now in a position to see the beauty of the dance. As a child, I lived to dance. As an adult, I could accept the fact that I loved to dance. When I felt warm, I walked across the studio and joined my classmates at the barre.

Eden Medina is Assistant Professor of Social Informatics at Indiana University.

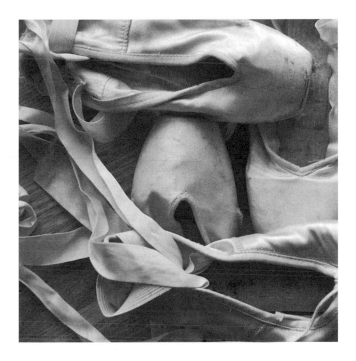

[Today] any objects or persons can be reasonably thought of in terms of disassembly and reassembly; no "natural" architectures constrain system design. . . . "Integrity" or "sincerity" of the Western self gives way to decision procedures and expert systems. . . . Human beings, like any other component or subsystem, must be localized in a system architecture whose basic modes of operation are probabilistic, statistical. No objects, spaces, or bodies are sacred in themselves; any component can be interfaced with any other if the proper standard, the proper code, can be constructed for processing signals in a common language. . . . The privileged pathology affecting all kinds of components in this universe is stress—communications breakdown. . . . The cyborg is a kind of disassembled and reassembled, postmodern collective and personal self.

—Donna J. Haraway, "The Cyborg Manifesto"

THE ELITE GLUCOMETER

Joseph Cevetello

Every morning the first thing I do is search my apartment for my blue case. In it is my Elite Glucometer, lancet, syringes, and other blood glucose testing paraphernalia. Carefully I open a test strip packet, insert it into my glucometer, load my lancet device with a sharp, new needle, search the tips of my fingers for a choice spot, and prick myself. I squeeze my finger until a tiny droplet of blood forms and hold the glucometer close until the vacuum pulls in the correct amount of blood.

The counter on my glucometer begins to count down time. It becomes my body's meter. I live by its metric. I might use the next sixty seconds to walk to the refrigerator to retrieve my insulin, or begin to make some coffee, or put my head down and think about going back to sleep. After sixty seconds, the meter displays my blood glucose level in milligrams per tenth of a liter of blood.

It is only recently that I have thought about how my meter, the first object I see every morning, has become me. Our interactions define my sense of who I am. My glucometer is credit card size—three inches times two inches, and is about one half-inch thick. It weighs about three ounces. The meter has no buttons or switches; it turns on only at the insertion of a test strip, which is about one inch long and one-quarter inch wide. At one end of the strip, an opening pulls blood into the testing plate. On its front is a small LCD display.

I have always been happy knowing that my meter is one of the most accurate on the market. I have been uninterested in how the meter determines my glucose level. The output is the event. I accept what my meter tells me.

Diabetes is all about control: control of blood sugars, control of what one eats and when one eats it, scheduled exercise, and regulation of insulin intake to food. However, there is no guarantee that even if you keep your disease "under control" its many side effects will not materialize. Despite a regimented life, you could still lose a limb or a kidney, become blind or impotent.

I was diagnosed with Insulin Dependent Diabetes in 1995. IDD is caused when the body attacks and kills off the insulin-producing beta cells in the pancreas. Insulin injections are required for all who have IDD. IDD increases the probability of heart disease fourfold, is the leading cause of kidney disease, limb amputations, blindness, and can lead to impotency. Left untreated, IDD would lead to death in about two years. When I was diagnosed I was fortunate to be living in Boston, the home of the Joslin Diabetes Center. It was at Joslin that I learned to care for myself and to be humble about my illness. While at the Joslin Clinic, I saw patients in wheel chairs who had lost a foot, others walking with IVs, and others with eye patches over one, sometimes two, eyes. I understood that diabetes was not something to be fooled with.

At the Joslin Clinic, I was introduced to the idea of "tight control." Tight control is the attempt to keep diabetic glucose levels as close to those of nondiabetics as possible. To stay on tight control I test my blood at least four times a day: in the morning, before lunch, before dinner, and before bedtime. On days when I exercise, I may test two times before vigorous activity to ensure my blood sugar is high enough and one time after I exercise to ensure that I have not gone too low. If I feel strange sometime during the day, I will test again.

What do I do with all this data? I write the data in my log book, in which I keep a tally of my glucose levels. The meter also stores my last thirty readings and can supply me with an average of these last thirty scores. As

I record the number in my log book, I project where I want to be throughout the remainder of the day, whether I can eat, how much I can eat, how much insulin I should inject, and whether I can exercise or must wait to get my sugars higher.

Usually, I come in at around 100 mg/dl (milligrams per deciliter)—the goal I have set for myself. If I meet this goal, give or take ten points, I feel a sense of accomplishment, a willingness to meet the day. If the read-out is much above 115 mg/dl, however, my mood changes abruptly. "A poor beginning," I say to myself, "What did I do? What on earth did I eat yesterday?" The next few minutes are spent reconstructing my last night's meals and insulin injections, adjusting my dose for the day, and thinking about what I can eat for breakfast.

I do not expect to be perfect, and I know there are times when things get out of control either because I ate too much or injected too little. Usually such readings do not bother me. But, "usually" is a big word in the world of "tight control."

There have been many times when I have thought I was low—when I even felt low—and my meter has told me the opposite and vice versa. Discrepancies of more than thirty points upset me. Sometimes I will remember a snack or lack of a snack, and that will explain it. Many times I can think of no good reason for the discrepancy. When my mental image of my physical self conflicts with my meter, I have a problem. Do I doubt myself, or do I doubt my meter? Seeking to maintain my sense of control, I test again.

My first reaction is to doubt the meter rather than myself even though I know that first and second meter readings usually differ by no more than five points. One would think that after all these years I would simply accept the first reading. I do not. I am unwilling to place absolute trust in my meter. I want to find fault in it, although I know it will always come up with two similar readings. The discrepancy between the reading and my expectation makes me redouble my efforts to remember what I could have forgotten, what I might have done wrong. Only when I remember do I feel in control once again.

My meter maintains my image of myself as a man able to take care of himself. It also defines me as a diseased person, one who needs the aid of objects to sustain my life. The meter concretizes my commitment to remaining healthy and communicates to others that I am different, somehow incomplete.

My interactions and dependency on my meter have made me realize that relationships between people and medical machinery are evolving. Perhaps, these new relationships will become so vital to our survival that, like my glucometer, they will seem intrinsic.

Projecting into the future, I can see two scenarios. In one, techno-clad humans live with ubiquitous computing, integrated into our homes, clothes, and bodies. I imagine data glasses receiving information about us from sensors buried deep within our bodies that could communicate a constant readout of blood glucose level. I, as the wearer, closely monitor myself and, at the appropriate time, communicate with my insulin delivery device to tell it to medicate me. In this scenario, I am in control of these devices, they do what I tell them to. In this fantasy, I am still a diseased person caring for myself.

In a second scenario, I live in a world of ubiquitous, body-based, clothing-based computing, but in this future, a small implantable device regulates my glucose levels and insulin needs. It operates autonomously. In this fantasy, I do not control my disease; my computer pancreas controls it for me. Manfred Clynes, a NASA scientist writing in the 1960s, defined a cyborg as a synergy between a machine and a human being that does not require any conscious thought on the part of the human.[1] In the second scenario, it is difficult for me to remember that I have diabetes. I have become, in Clynes's terms, a cyborg. I wonder how my interactions with my meter may be a harbinger of the nascent stages of a cyborgian relationship.

The Austrian poet, Rainer Maria Rilke, said: "The future enters into us, in order to be transformed in us long before it happens."[2] I find my blue case and take out my meter, blood glucose testing strips, lancet device, syringe. Carefully I open a test strip packet, insert it into glucometer, load up my lancet device with a sharp, new needle, search the tips of my fingers for a choice spot, and prick myself. I squeeze my finger until a tiny droplet of blood forms and hold the glucometer close until the vacuum pulls in the correct amount of blood. As the meter counts down, I begin to prepare my shot and wait for my meter to tell me what to do.

Joseph Cevetello received his doctorate from Harvard University School of Education and is a specialist in e-learning design and technology use in adult learning.

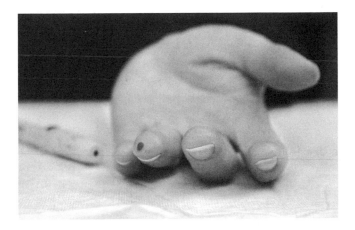

Strange indeed is the encounter with the other. . . . Confronting the foreigner whom I reject and with whom at the same time I identify . . . I lose my composure. I feel "lost," "indistinct," "hazy." . . . [Yet] the foreigner is within us. And when we flee from or struggle against the foreigner, we are fighting our unconscious. . . . Delicately, analytically, . . . [we must be taught] how to detect foreignness in ourselves. . . .

By recognizing our uncanny strangeness we shall neither suffer from it nor enjoy it from the outside. The foreigner is within me, hence we are all foreigners. If I am a foreigner, there are no foreigners.

—Julia Kristeva, *Strangers to Ourselves*

THE YELLOW RAINCOAT

Matthew Belmonte

Even in primary school I was preoccupied with the idea of protection from an unpredictable world. Protection often came in the form of a glaringly bright, yellow raincoat that kept me dry on rainy days on my way to school. A thoroughly synthetic creation made of rubberized polyester, it would have been difficult to imagine anything less natural. It would be difficult to imagine an artifact that more embodies the tension between myself and my environment. More than its function of keeping rain out, however, it represented my fear of letting anything in—people most of all.

People were the most unpredictable elements of my world; unlike other objects they were more than the sum of the forces acting on them. The human factor was a constant irritant for a budding Laplacian like me. Where a person was involved, one could never be assured of predicting the output, even if all the inputs were known. My wish back then was that I could be the human analog of the neutrino I had read about in science articles: a particle that moved effortlessly through the world, almost never interacting. On the playground, while the other three-year-olds competed for the swings and the slide, I paced along the fence, studying the ground and identifying minerals in the rocks that I found. Rocks, unlike people, were safe.

Wrapped around and covering me, the raincoat represented my mother's triumph over my own will, and persistently reminded me of my dependence on her. In a fundamental way that I didn't consciously acknowledge, the coat came to represent my mother, and I loved and resented it as I loved and resented her. A fear of death, of being smothered and negated, drives us to separate ourselves from our parents. And a fear of life, of being responsible for ourselves in an indifferent world, brings us back to seek their protection. These conflicting denials of death and of life were attached to the coat: it made me impermeable to the assaults of the outside world, yet it defined me in a way that prevented me from being myself.

In solitude I slipped between the horns of this dilemma. When I was alone, there was neither the threat of attention from other people, nor the demand to submit to the decisions of my parents. The defeat of my will that was signaled by the yellow coat could be replayed as a victory, if I were the one who chose it. Walking alone through a downpour, I was immersed in the outside world's flood yet insulated from it. It was thrilling to feel the pressure of the rain and to see it roll off me and leave me dry. It was as if I were marveling at some alien world and knew that a spacesuit was all that separated me from its deadly atmosphere. Alone in the rain, I was master of my own actions and of my surroundings.

I believe that my childhood sensitivity to the boundary between self and external world led me in my adult life to study people with autism, whose central, daily challenge is the work of imposing internal narrative flow on a deluge of external sensory inputs. Ironically, when I was in primary school I never felt much empathy for my autistic older brother. Now as I look back I see both science and autism are compulsions to order, which differ only in their degrees of abstraction. I now feel that the same set of genetic biases that gave my brother autism gave me just enough of a desperation for order to make me a scientist, and indeed, a student of autism—enough to be driven by the same sense of impending chaos that drives my brother, yet I'm not as overwhelmed by it. I often consider how similar he and I are, and how I so easily could have been him, or he me.

So it was this shared desperation for order that drove me into science, and later into the craft of fiction. Like my old raincoat, science and art enable me to immerse myself in nature's order while they insulate me from nature's chaos. As scientists we invent perfect models in which phenomena are supposed to be mathematically tractable; the human construction of science is full of ideal gases, incompressible fluids, frictionless surfaces, and blackbody radiators. Similarly, as artists we filter the complexities of real life into representative texts in which distinct characters are involved in coherent plots evincing meaningful themes. Treating life as theater and inventing purpose and order, I keep chaos, meaninglessness, and death at bay. My theoretical and narrative constructions in science and art are the same sort of protective gear as the impermeable coat that I once wore to primary school; they hold nature at arm's length, close enough so that I can make some sense of it, but far enough so that I won't be overwhelmed.

My work has taught me that this notion of protection goes a long way toward explaining how people construct theories to gain a sense of control over their surroundings. Then they behave in ways to reinforce these theories. People with autism share the "normal" desire to control their surroundings. What differs for them is the intensity with which these surroundings impinge. Abnormal neural connections within autistic brains may lead to abnormal perception, increasing the salience of individual events but undermining the ability to connect these pieces of life into more integrated and abstract representations.

I made understanding the experience of such a fragmented perceptual world the center of my work. To proceed, I imagine life as a film being screened by an incompetent projectionist. Perhaps the volume is so high that none of the dialogue can be heard above the hiss of noise, or perhaps the aperture setting causes one bright corner of the picture to drown out all the rest. However, if I can rewind the film and play it again and again, I can gather a bit more information each time I watch it. My aspiration is to understand all of it.

The rigid and repetitive behaviors of people with autism begin to make sense when we consider them as the normal reaction of a human mind to a very abnormal sensory environment, rather than as direct symptoms of an illness. Autistic symptoms are what a person does in order to force a chaotic world to follow a predictable script. We are all trying to impose a narrative order on what may seem a fundamentally chaotic world. The difference in autism is that there is more chaos to be controlled. In this regard, the study of autism can tell us a great deal about humanity in general and how psychological distress can be explained as a rational, if extreme, reaction to a world gone awry.

On a stereotypically rainy English day, I still enjoy a ramble through the countryside. Trudging through the rain helps me collect my thoughts about science and life. As I squelch along footpaths, I consider that each raindrop is an observation in itself, and I marvel at the task of comprehending the storm without drowning in it.

Matthew Belmonte studied the neurobiology of autism at the University of Cambridge and is now at Cornell University in the Department of Human Development.

For the clock is not merely a means of keeping track of the hours, but of synchronizing the actions of men. . . . The bells of the clock tower almost defined urban existence. Time-keeping passed into time-serving and time-accounting and time-rationing. . . . The clock, moreover, is a piece of power-machinery whose "product" is seconds and minutes: by its essential nature it dissociated time from human events and helped create the belief in an independent world of mathematically measurable sequences: the special world of science. There is relatively little foundation for this belief in common human experience: throughout the year the days are of uneven duration, and not merely does the relation between day and night steadily change. . . . In terms of the human organism itself, mechanical time is even more foreign: while human life has regularities of its own, the beat of the pulse, the breathing of the lungs, these change from hour to hour with mood and action, and in the longer span of days, time is measured not by the calendar but by the events that occupy it. . . . To become "as regular as clock-work" was the bourgeois ideal, and to own a watch was for long a definite symbol of success. . . . By now Western peoples are so thoroughly regimented by the clock that it is "second nature" and they look upon its observance as a fact of nature.

—Lewis Mumford, *Technics and Civilization*

THE DATEBOOK

Michelle Hlubinka

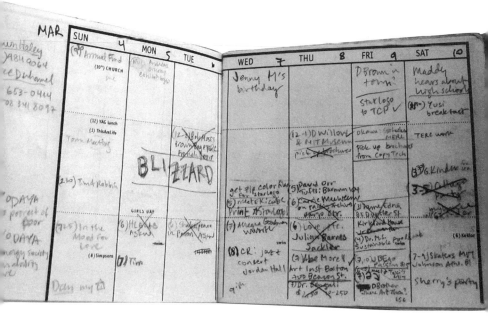

Benjamin Franklin aspired to be a person of high moral virtue and used time management technology to try to become that person. He created a planner and explained his effort, writing, "I conceiv'd the bold and arduous project of arriving at moral perfection. I wish'd to live without committing any fault at any time; I would conquer all that either natural inclination, custom, or company might lead me into."[1]

A page of his planner reads:

THE MORNING.
Question. What good shall I do this day?
5 – 6 – 7
Rise, wash, and address Powerful Goodness!
Contrive day's business, and take the resolution of the day;
prosecute the present study, and breakfast.
8 – 9 – 10 – 11
Work

NOON.
12 – 1
Read, or overlook my accounts, and dine.
2 – 3 – 4 – 5
Work

EVENING.
Question. What good have I done today?
6 – 7 – 8 – 9
Put things in their places. Supper.
Music or diversion, or conversation. Examination of the day.

NIGHT.
10 – 11 – 12 – 1 – 2 – 3 – 4
Sleep.

In childhood, we experience the passage of time as undifferentiated flow, marked by naptimes, meals, sunsets, and the familiar jingle of a favorite daily cartoon. Other people take charge of our schedules. As children we are inducted into the responsibility of managing our own time when we are taught to read the face of a watch.

My own induction into adult time took the form of a wind-up Mickey Mouse watch I received on a family trip when I was four. Before I owned the watch, I had all the time in the world. When time was packaged into a wrist-sized, mechanically driven object that I carried with me, I could watch time pass. Mickey's arms flailed across the hours. Time materialized and evaporated. Having the watch, I entered a society not just of time-keepers, but time-managers. And I became good at it, perhaps too good at it.

And then, one spring, I lost my datebook. I felt as though I had lost my life.

My memory of all I did and planned to do from January to May 2003 vanished, along with the physical form that contained it. Within the ratty, dog-eared pages of my datebook I had inscribed the talks and lectures that caught my eye, the art-house movies I enjoyed, the meetings that filled my days, the friends I met for lunch and dinner. My datebook and its events had their own esoteric language. Familiar venues, organizations, and individuals were noted in tiny writing and abbreviations that only I could decipher. Sometimes I would scrawl dozens of coded commitments into rectangles not more than an inch square. My datebook enabled me to weave a matrix of possibility: I would often note three concurrent events that sounded equally enticing, and at the last minute my whims would direct me to one of them or to cross them all off my list.

Now that my datebook is gone, I still wonder what commitments I may not have honored in the weeks after its disappearance. I think of my lost datebook as an external information organ—a piece of my brain made out of paper instead of cells. Knowing it was nearby helped me relax. Like an early computer toy that needed a storage cartridge to expand its capacity, I needed my thickly layered and coded pages. My sense of self as cyborg sometimes bothers me. I envy my friend Mieke who does not own a calendar. She divides her time between weekdays (these begin full with work) and weekends (these begin empty and fill up with friends or research). Mieke can carry her life in her head. She fills her *life* not her datebook with events.

The style of my encoded datebook also contrasts with the transparent aesthetic of my friend Ginger. In Ginger's datebook, each double-page spread covers a week. Ginger's datebook is color coded, each color corresonding to an aspect of Ginger's life: personal (red), work (blue), and nonwork professional interest (green). Ginger says that her datebook reveals a great deal about her bulimic past: "It feels irritating when I don't have the [right] color pen. . . . I think it relates to my need for control—which goes back to my eating disorder, actually. I need to have a sense of what things are boxed where. . . . I make the boxes and I put it in order and I know how much I can fit into one day." Ginger inks in her appointments with space around them. She has a horror of being late ("in my head, it feels like I am going to end up disappointing somebody") and protects herself against being late by never having a time slot that physically abuts another. One day, an overlapping pair of rectangles did appear in her datebook. It made her anxious: "If I can't do everything then I am going to end up being rejected. People are not going to like me, so I'm going to have to fit it all in. The overlap—that's anxiety-provoking." For Ginger, being on time is a way to not draw attention to herself. "As an undergraduate, for instance, when I was going to be five minutes late for a class, I would prefer not to go to class at all, even if it was a three-hour class, because I was so afraid that if I showed up late people would stare at me and think I was fat. Like, I'd walk in the door and they'd think, "Oh there's that fat girl. She's so late. What's wrong with her? For me, being perfect meant being on time."

When she had an eating disorder, Ginger recorded her food consumption for each day. Every evening she would go through the list and highlight the things that she felt she should not have eaten. Now, she uses the datebook to keep track of her exercise. Each workout is marked with a cheerful cartoon sticker. She says, "When I have jogged and done calisthenics I give myself two stickers so I can feel good about myself. I have a very visual picture so I can reward myself for exercising." Ginger cares too much about the image others have of her. But with her calendar, "My audience is myself . . . a lot of these devices are to make me happy."

Like Ginger, who used her calendar to track her goals of punctuality, healthy eating, and frequent exercise, Benjamin Franklin constructed a system that required a daily chart of what he would accomplish and when. A second chart tracked Franklin's adherence to thirteen core virtues (temperance, silence, order, resolution, frugality, industry, sincerity, justice, moderation, cleanliness, tranquility, chastity, and humility). Franklin placed a check every time he violated a virtue on the chart. Each week Franklin concentrated on a different virtue and sought to have its row free of checkmarks.

Franklin's datebook helped him to visualize his progress toward becoming the person he wanted to be. In 1980, Hyrum Smith read Franklin's autobiography and took it as a model to invent the most popular planning diary in America that stressed priorities and daily tasks. Lars, a computer science graduate student, doesn't use the commercial system, but his computerized priority list is nonetheless a direct descendent of Franklin's chart:

> Twelve years ago I wrote everything on a priority list. So when I got a computer, I put all those scraps of paper on it. It is a precise description of everything I want to do in my life. . . . It is 100 or 200 pages long on my computer. Every day I went through the whole list and I took the most important things out of the list and printed it out. It would be one or two pages at most.

When Lars was self-employed, he didn't need a calendar. His everyday printout of his highest priorities was enough to keep him on track. Now, in graduate school, Lars uses an armory of interconnected devices to map his life. Some keep private time that is concealed (Microsoft Outlook on his home PC, a paper printout of the Outlook calendar, a Timex watch that downloads appointments from Outlook and will beep to remind him when one is coming up); others mark public time that is revealed (a UNIX-based calendar that he shares with his workgroup, a pager and cellular phone that is linked to this work calendar). Always crucial to Lars is the distinction between public and private calendars: "My calendar on my PC is my private thing. . . . And actually I am very reluctant to show this paper [he holds up the printout] to other people. It's a very private thing." I realize that Lars's private computer calendar is no less an intimate projection of self than my color-coded datebook. But whereas I want to keep all the choices I did not make as a permanent record of paths not taken, Lars wants his computer calendar to be a pristine reflection of his life as lived:

If I don't go to something, on a computer I just delete it and it is clean. If you do this with a paper calendar . . . it would perhaps make me feel bad, because I would see how many things I wanted to do and did not do. . . . The past is not clean. In a computer calendar it is easy to forget: you erase, it is gone. You can clean your past very easily with a computer calendar, as you cannot with a paper calendar.

Since losing my datebook, this distinction between paper and computation has been much on my mind. When I found myself without a datebook, I began using iCal, a piece of software that came with my Apple PowerBook. I needed a place to jot down my commitments, and it was readily available. Instead of carrying around my schedule with me all the time, I look each morning and try to remember my appointments. If someone asks to meet up for tea, they have to wait for me to check the master iCal schedule when I get back to my computer. I find I make fewer appointments. Moving my datebook to the computer is an experiment. I would say that it is not going well. My new system leaves me destabilized. I still write down all the talks, lectures, and movies that catch my fancy, but I have not yet decided what to do with those I *don't* attend. If I leave them in my schedule, I won't be able to differentiate what I did and did not do. But I'm not like Lars, comforted by a pristine record. If I delete the options I did not take, I mourn that I have lost a record of what interested me in a given week; I lose the shape of what happened in my community. I like to think that anyone could open up my lost paper datebook and see what kind of person I am. I imagine my runaway, tattered paper pages wandering around Cambridge, Massachusetts, being picked up by strangers who try to decipher who I might be, perhaps attending some of the events I had noted for April and May and looking around for me, the curious owner.

Michelle Hlubinka, M.S., Ed.M, an educator, designer, illustrator, and storyteller, is education manager at Zeum, an arts and technology museum in San Francisco.

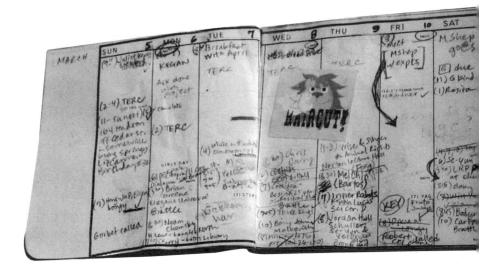

Ultimately, computers demonstrate that we cannot only project and win back this one universe, but that we can do the same with as many as we want. In short: our epistemological problem, and therefore our existential problem, is whether everything, including ourselves, may have to be understood as a digital apparition. . . .

[But] it is not enough to acknowledge that the "self" is a node of criss-crossing virtualities, an iceberg swimming in the sea of the unconscious, or a computation that leaps across neuro-synapses: we also have to act accordingly. The alternative worlds emerging from the computers are a transformation of this understanding into agency. . . .

[Computers] thus realize . . . alternative worlds and thereby themselves. . . . [They] are apparatuses for the realization of inner-human, inter-human, and trans-human possibilities, thanks to exact calculatory thought. . . . We are no longer the objects of a given objective world, but projects of alternative worlds. . . . We grow up. We know that dream.

—Vilém Flusser, "Digital Apparition"

MY LAPTOP

Annalee Newitz

My laptop computer is irreplaceable, and not just for all the usual reasons. It's practically a brain prosthesis. Sometimes I find myself unable to complete a thought without cracking it open and accessing a file of old notes, or hopping online and Googling a fact or two.

Besides, I love it. I would recognize the feel of its keyboard under my fingers in a darkened room. I have worn two shiny spots on it where the palms of my hands rest when I'm not typing. I carried it on my back all over England, Cuba, Canada, and the United States. When I use it in bed, I remember to keep the blankets from covering its vents so it doesn't overheat. I've taken it completely apart, upgraded its RAM, and replaced its original operating system with Linux. It doesn't just belong to me; I also belong to it.

I'm hardly alone in my infatuation. When I was fifteen, my friends and I would often stay up late into the night, chatting online over a multiuser chat system called WizNet. Using online aliases, we spent hours talking about science fiction, movies, computers, and sex. My alias was Shockwave Rider, a reference to a science fiction novel about some guy who hacked phone systems. I was the only girl in the group, although you wouldn't have known it. Like everybody else, I was just a command line full of glowing green letters.

Some of the WizNet denizens knew who I was "in real life." Eagle was my best friend. Punk Tofu had met me at AppleFest, a convention where Steve Jobs unveiled the first Macintosh. Sauron, Splat, and Ectoplasm went to my high school. When we weren't studying algebra or tenth grade English, we were scarfing down donuts and playing video games together at Eagle's or Sauron's house. It was during one of those sugar-fueled sessions that I first heard the name Gonif. He was a culture hero on WizNet, a cracker who'd broken the copy protection on MacPaint and dozens of other programs for the Apple, thus liberating the programs and allowing all of us to use them for free.

Gonif had also supposedly been the victim of a cruel prank. Someone had sent him an e-mail pretending to be a girl who wanted to go out with him. When Gonif showed up for the "date," he found a bunch of guys laughing. Supposedly he only showed his face on WizNet now under an assumed name.

A few days later, I met someone on WizNet who called himself Josh. He was a better writer than anyone I'd ever met on WizNet, and for the next few weeks, we'd occasionally meet on the electronic bulletin boards late at night to talk about books and our teenage life philosophies. We'd go into a private chat area, where nobody else could read our words. It seemed like I had no thought that he couldn't understand.

One night, after a long discussion about ethics with him, I told him via type, "I love your mind." Josh revealed to me soon after that he was Gonif. He was the elusive cracker I'd been hearing about, the guy who was fiendish enough to crack MacPaint and yet still so romantic that he'd fallen for the fake e-mail trick. At that moment, when I learned his true name, I fell passionately in love with him.

His body was a green light on a Kaypro screen and the feel of slightly concave keys nestled in a brushed stainless steel tray. His breath was the sound of a fan cooling the CPU. I heard his voice in the sound of my modem; I saw the most beautiful parts of him in the shape of his sentences as they emerged out of the ether and entered my mind whole. I loved him for what he could do with language and computers.

The whole affair lasted only one summer, but those late nights with him and my computer remain in my memory forever. Almost fifteen years later, I managed to track him down again and we exchanged some e-mails. He was still the same thoughtful romantic. Seeing his words on my monitor reminded me forcefully of what had originally drawn me to him. It had nothing to do with physical prowess or money. It was his mind and the things he created with it.

Of course my affection for Gonif determined my relationship with my computer. How could it not? To this day, every time I boot up my machine, I see a shadow of him flicker past.

Harvard professor of clinical psychiatry John Ratey says that because our brains link ideas together in memory, we are particularly well-suited to the act of suffusing an object with emotional value.[1] If someone I love gives me a portable wireless device, it's likely the device will remind me of that person. It's easy to see how this would work in a purely personal context: I have my own unique sets of associations, ,and therefore one could write off my peculiar passion for computers as a simple trick of fate. It just so happened that I had early romantic experiences with machines, and so computers make me think of love.

But how do you explain all the other people who adore their computers? Several hundred thousand people visit the infamous "news for nerds" computer-lover Web site <http://slashdot.org> every day. How does an entire subset of a society learn to associate feelings of pleasure with the same kind of object?

I think back to a soulful conversation I once had with Richard Stallman, an MIT computer science researcher and activist who began a largely geek-centered movement known as the free software movement. Through this movement, he advocates that people build software that they openly share. This doesn't mean they should give it away for no money, although often they do.

For Stallman, creating free software gives you the liberty to modify it as you choose, and to use it however you like. For Stallman, free software flows from community to community in bonds of sharing.

Stallman spends nearly all his waking hours on the computer, building software and communicating with fellow activists in the Free Software Foundation. When he was a young man in the early 1970s, he invented a powerful tool called GNU/Emacs for building software. It was quite an accomplishment and earned Stallman a permanent position at MIT as well as countless awards. Linus Torvalds, inventor of the popular Linux operating system, was so inspired by Stallman's work that he used GNU to build the kernel of software at the heart of Linux.

To me, Stallman is a romantic. He told me he dreams that one day the free software idea will affect all society. In that world, he said, he would find his perfect love. She would share herself with him the way he has shared his ideas and tools with so many people. And her love would not be jealous or selfish: she would give to him in a perfect relationship of reciprocity. Stallman loves his computers because in them he sees a web of altruistic social relationships. He doesn't spend all his time at the keyboard to avoid other people. He does it because one day, he wants to fall in love again.

Annalee Newitz publishes regularly on technology in national magazines and newspapers.

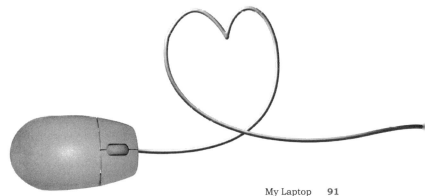

In the classical period, it is futile to try to distinguish physical therapeutics from psychological medications, for the simple reason that psychology did not exist. When the consumption of bitters was prescribed, for example, it was not a question of physical treatment, since it was the soul as well as the body that was to be scoured; when the simple life of a laborer was prescribed for a melancholic, when the comedy of his delirium was acted out before him, this was not a psychological intervention, since the movement of the spirits in the nerves, the density of the humors were principally involved. But in the first case, we are dealing with an art of *the transformation of qualities,* a technique in which the essence of madness is taken as nature, and as disease; in the second, we are dealing with an art of discourse, and of *the restitution of truth,* in which madness is significant as unreason.

—Michel Foucault, *Madness and Civilization*

BLUE CHEER

Gail Wight

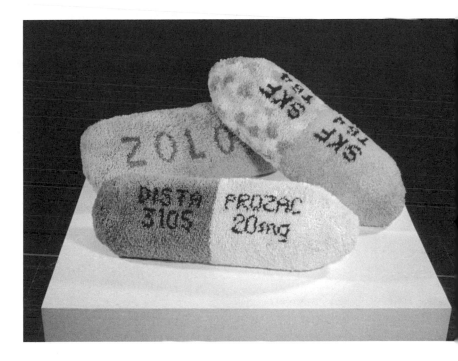

At the back of my top dresser drawer is a black calfskin wallet with a plaid interior, its fine stitching torn where the leather itself started to tear. Finally too worn to be used, the wallet holds something more important, my last tab of Ludiomil.

A tab of Ludiomil is not something one saves for a rainy day. It wouldn't do anything on its own, like a faded hit of acid might, or a bit of shriveled mushroom, or a nip bottle rescued from some uncertain vacation. It needs at least thirteen others of its kind to be effective—doled out steadily, one per day—until two weeks later, swallowing the fourteenth tab, one might feel something a little different. More likely, one would sense the absence of a familiar feeling.

This tiny hard oval of robin's egg blue is a placeholder, a minuscule iconic bombshell. Just looking at it brings a whole world home to roost. With more than 4,000 legitimate variations and an untold number of underground permutations, the ability of pharmaceuticals to provoke a steamer trunk-full of associations, opinions, and emotions, is new only in its industrial and technicolor nature.

Drugs have been around. These new drugs, though, bring a cartoonish sheen to the shriveled earthtones of innumerable entheogenic plants and animals. Their images spice up magazine covers, pop music CDs, and movie posters. Like the glowing virgin's halo in medieval iconography, the evocative power of pharmaceuticals is bound up in their appearance. With growing regularity, the image of pills inhabit everyday marketing, but I especially love to find them in the aerial landscape of contemporary art: the pill-popping geisha of Ridley Scott's sky-high illuminated and animated billboards; Fred Tomaselli's pill paintings, hovering somewhere between wallpaper and psychedelic string theory; Laura Splan's giant cozy capsules of Prozac, Zoloft, and Thorazine revisioned as huge needlepoint pillows for a weary head. Their ubiquitous presence provokes a rash of reactions from those who will never need them, as well as those who will, and those already acting on their need.

From an aesthetic point of view, one could be nostalgic about these tiny icons. What happened to the silver and cloisonné inlay of an antique opium pipe, or the elaborate artwork on a thumbnail-sized canvas of LSD? I feel old when I think about drugs delivering themselves from an implanted digital pump, or the bland and murky translucence of a skin patch. The institutionalized prescription of Brave New World's "soma" is part of our culturally shared fear of the future. I like my drugs colorful, aesthetically inspiring, exhilarating the imagination through the sheer force of their physical beauty—and I know that I'm not alone in this. There's a reason the American landscape is littered with glowing neon signs that say simply: DRUGS. These signs are visually seductive gateways; it follows that drugs should present themselves in vibrant and tantalizing packages. I deplore the timid pastel plastics of birth control blister packs, for instance. I want to see a visual onomatopoeia, a symbolic poetic linkage of image and impact. Still, digital timers, patches, and implants have their appeal: comedian Betsy Salkind says, "I've never smoked, but I sure do love the patch."

But my lone little Ludiomil *is* beautiful, a pale blue speck of sky. It's sweet and tiny and powder pastel, not so devoid of aesthetic pleasures after all. It reminds me that the pink and blue of infant codification was reversed in Victorian times, pink being for boys and blue for girls. This gives me a fierce feeling of ownership for my blue dot, a feminist protectorate. It helped me escape from whatever might have spawned my depression: a hereditary noose, an intractable post-partum guest, a late capitalist malaise in a post-punk dress code.

While that depression was likely a collaborative effort, borrowing something from each of the above, I'm most comfortable blaming my ancestors for that neurological genetic ambush, and contemporary medical theory encourages me to do so. With promotional brochures in hand, I headed to the pharmacist to stock up against further attacks from my relatives. To be sure that one's relatives are the root cause one needs evidence, of course. This can be awkward or impossible and pointless in a family where mental health is a nontopic. But once my sleuthing had revealed a ghost or two, or three, every relative living and dead began to look a little guilty to me. But this was all good news: protocol dictated that producing such a relative for the doctor would be key to an instant prescription.

And there was my Uncle Bob. Uncle Bob was easy to beat at Monopoly, but I chalked this up to adult benevolence. He would play for hours, until my mother intervened, and I adored this about him. He drove a car, and I loved the way he did this as well. He never went over fifteen miles per hour and traveled mainly on the shoulder, so cars would screech around us, honking. A trip to the next town to get a piece of sheet music meant hours away from my home, meandering through Connecticut farmland. I don't remember that he played any instruments, but he loved sheet music and would bring favorite pieces to my mother as gifts. He would bring her other gifts as well, appliances wrapped in their store boxes like new, but always used and often still dirty from use, like a greasy electric griddle he gave her one year for her birthday. He was adorable and odd, and sometimes he laughed "too hard." As a child, I assumed these quirks were just part of the adult condition, until I was older and had an adult condition of my own. Suddenly, I wanted to know more about Uncle Bob.

Along with a distressingly large percentage of the population, I'd been raised with a deep distrust of therapy. Wasn't that where they changed who you actually *are*? Did I really want my true self to be changed? Well, yes. I had reached the point where my self had become an anchor, one that would drag me to the bottom of the Charles River if I didn't transform it somehow. I had become fond of my depression (though I didn't have that name for it yet), but it was essential to attempt to shake it, just the same. I was in art school at the time, and in the course of free therapy offered at my school, I was asked to examine the past for clues to my desire to die young.

And so I looked back; it didn't seem that interesting. In grade school, my first favorite pop song—the one that separated me from the influences of my older brother and sisters—was Alice Cooper's "Dead Babies." I wore a lot of black, preferably original items from the late 1800s, and was convinced that I could see my skeleton, whenever I looked in a mirror. If there had been Goth, I would have been Goth—made up carefully, daily, to resemble the newly deceased. My high school drawings were landscapes of New England cemeteries, gravestones, funereal flower arrangements, and dead roses. If I were a teenager today, with this same profile, I would probably be expelled as a potential shooter. But it was the 1970s, and suicidal tendencies were a normalized part of youthful rock and roll. So none of this typical teenage angst added up to much of an offering for a therapist.

After two years of conversation focused on good living strategies, my depression remained, entrenched and intact. At some point, my relatives, my family profile, the guilty ghosts, made their appearance; my therapist suggested medication. I thought at the time that she had finally become bored with me, and maybe she was, but she was also saving my life.

The first drug I tried, Meritol, took an interminable two weeks to hit me, as the tiny tabs added up their punch. And then quickly, I became intolerable. Happy all the time, I laughed "too hard" at anything, especially the tragic. But I was happy. I was ecstatic, really, full of energy and able to focus. I assumed that what I was feeling inside was what I saw from the outside, when I met "normal" people, and I began to see depressed people everywhere. This was good technology.

One night, just after midnight, the phone woke me up. It was the doctor who was feeding me my experimental antidepressants. He wanted to know if I'd taken my medication that day and if I had much left before the next refill. His voice was urgent. He ordered me to discard any Meritol I had left and stop taking any other meds immediately, and then he suggested I call in the morning for an appointment. People were having heart attacks. Two people had died. Meritol was flawed technology.

I don't remember much about my second medication. In fact, I don't remember much about the many weeks that I took it, because I was asleep most of the time, dreamless, lethargic, and deeply depressed. I managed to drag myself to work and to a few classes, but was never awake long enough beyond that to make any phone calls. My husband intervened on my behalf.

Next came Ludiomil. A happy middle ground. I didn't want to die and I didn't laugh too hard. Happy was again the word, but this time I didn't feel drugged. In fact, I rarely thought about the fact that I was medicated and began to live a life where I was comfortable in my skin for the first time, confident enough to just be. I took Ludiomil for three years, and then, when I started to feel medicated for the first time, I tapered off without consulting my doctor. Two years later, with the nation deep into Prozac culture, it was much easier to get a prescription.

We are all subject to the concepts that drive the world during our time here. My uncle survived many changes in cultural conceptions of health, medicine, and madness. He escaped a lifetime of institutionalization, survived electroshock therapy and a lobotomy, and managed a large degree of independence. He avoided other technologies—some still in use in the 1940s, others long consigned to the dustbin of medical history: forced sterilization, ice baths, and insulin shock treatments, all still popular then. Just a century or so earlier, doctors might have sentenced him to chains, the Bedlam Crib, or the bleeding and purging of Benjamin Rush's "tranquilizer chair." I suppose I feel fortunate to live in an age of heavy reliance on drugs, but I'm also aware that no one can really tell me if there's still some price to be paid for my days of medication.

If genetics are in fact a key player in the severe mood swings that plague my family (I have little doubt of this, given the grocery list of familial suicides and self-medicators), then I'm a guilty link in the chain. I worry about the mental health of possible grandchildren in my future. I wonder about shifts in medical technologies and cultural ideologies. Maybe the future will be free of Prozac, and something new and considerably less barbaric in its own right will take its place, or eradicate its perceived need forever. Maybe, in the future, tableting machines will send chills down people's spines.

Last week I went through my dresser, filling a few paper shopping bags with clothes from another lifetime, sweaters and T-shirts and a few old skirts that would find their way to Goodwill. I held the black wallet, so soft and emaciated. It still smelled a little of leather, mixed with indiscernible and ghostly perfumes. It was useless, a few holes along the edges, the zipper for the change compartment completely ravaged. But it had a bigger task. It was home to the lone Ludiomil, my own currency, my old ticket stub to happiness, my golden treasure, my blue cheer.

Gail Wight is Associate Professor of Experimental Media Art at Stanford University.

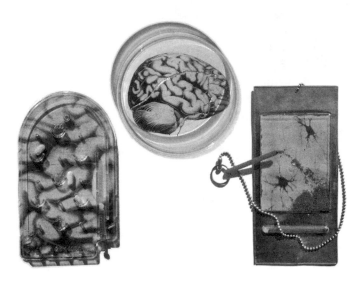

Objects of History
and Exchange

Functional perfection exercises a cold seduction, the functional satisfaction of a demonstration and an algebra. It has nothing to do with pleasure, with beauty (or horror), whose nature is conversely to rescue us from the demands of rationality and to plunge us once more into an absolute childhood (not into an ideal transparency, but into the illegible ambivalence of desire). . . .

All possible valences of an object, all its ambivalence, which cannot be reduced to any model, are reduced by design to two rational components, two general models—utility and the aesthetic—which design isolates and artificially opposes to one another. . . . But this artificial separation then permits evoking their reunification as an ideal scheme. Utility is separated from the aesthetic, they are *named* separately (for neither has any reality other than *being named separately*), then they are ideally reunited and all contradictions are resolved by this magical operation. Now, the two equally arbitrary agencies exist only to mislead.

—Jean Baudrillard, "Design and Environment or How Political Economy Escalates into Cyberblitz"

THE RADIO

Julian Beinart

The waterless coastline stretches thousands of miles, from just north of Cape Town all the way to Angola. I grew up in a small town at the southern tip of this desert and was a child when German submarines torpedoed Allied convoys and left survivors to waste away on this Skeleton Coast. My town was a hot and dull center for wheat farmers. The tallest building was the Dutch Reformed Church, an Afrikaner Gothic steeple, to which white dressed-up farmers' kids would march on Sunday mornings. My family belonged to the synagogue across the mud of a river, in an out-of-the-way place where its low, quasi-Ottoman façade faced no one. Colored people cleaned our house, drove my father's trucks, got drunk on Saturday mornings, and lived somewhere I did not know.

Later, when I was a sophomore in architecture school, I tried to do a measured drawing of the church for a class assignment. It was the only building in the town that seemed to merit my work. But I never was able to finish the drawing. The church was too big to measure, and somehow it stood outside me. It was a totally isolated and commanding thing, never to be messed with, never to be modified, never to change, and never to be entered by the likes of me, or, as I later understood, by all those colored Christians.

In many ways the Church fitted much of the dogma of the architecture I was taught. We never questioned client power or community access or social meaning in buildings. Our designed objects were to be seen on their own in space and to remain unaltered over time. We had the benighted obligation to innovate culture, a culture produced by Western heroes working for people like themselves. Our ideal was to have Palladio's clients, princes with whom we could act out our professional narcissism.

Years later I was in South Africa again, now with graduate degrees from American universities and a sense of obligation to spread their wisdom. But to whom? The universities were segregated; increasingly uneasy, I taught basic design to freshmen, based on what I had learned at MIT from Gyorgy Kepes, who in turn had brought his version of the famous Bauhaus Vorkurs from Europe. The exercises of this fundamental course were meant to reduce students' reliance on past visual knowledge and to force them to deal with a formal language of vision completely new to them. The new language was abstract and universal, implying that it could be as international, yet as removed from local culture as Esperanto. In what Baudrillard refers to as this universal semantization of the environment, visibility was controlled.[1]

Soon after, in the early 1960s, I remember how shocked I was when I saw something I had not noticed before. Walking down a street in the middle of Durban, South Africa's most racially mixed city, I passed a boy carrying a wooden transistor radio. It was about six inches long and two inches wide, with a wooden handle and a hinged wooden dowel antenna about two feet long tapered to a small knob at its end. On the top of its body, one of three square wooden buttons was pressed down. A slit of broken glass covered a rectangular dial behind which was a piece of an old paper calendar numbered one to twelve. A red pointer was stuck on three; it could never move. Although it looked like a Braun transistor radio, this object never produced sound. I asked the boy about it and he said: "It can't play music, but I sing when I carry it. One day I'll have a real one."

From that time, quite suddenly, I began to see objects that had been invisible to me before. There were all kinds of wire bicycles, some of twisted soft metal, others shaped out of thin steel with yellow frames, red beaded tires, blue handles, and pedals. A friend sent me a three-by-two-foot black bicycle from Zambia, which had a movable front wheel. It had, so he said, been made by a boy to get himself a job in a bicycle repair shop.

Everywhere there were objects of emulation and imagination. Often they were copies of sophisticated machines now made by hand out of recycled, thrown away material: Honda motorcycles made from panels of sheet tin taken from Castle beer cans; a dark green Isuzu Trooper 4×4 made out of a single piece of wood; wire Volkswagen Beetles with engine covers that lifted up; a snout-pointed fighter plane with a South African flag on its rudder; a large helicopter made of wire with a working AM radio in its belly. In the mute transistor radio family, there were silent wooden Sony cell phones useful only for dreamed conversations.

Cheaply available, highly visible, and linguistically subtle, material from products carrying popular brand names and out-of-context messages (Coca-Cola, Sprite, and Fanta, among others) adorned tin lunch pails, cloth jockey caps, miniature delivery trucks, and almost everything else. Recently I bought a three-foot-long pantechnicon in New York. Made in Abidjan of Nestle coffee can metal, it repeatedly says:

> *Nescafé est un pur café soluble, fabriqué avec des grains de Robusta de Cote d'Ivoire, soigneusement selectionnés puis traité pour votre plus grand plaisir.*

And on an elegant racing bicycle from Cameroon there are small-type messages about "milk for baby's growth" and "just add water."

I have puzzled over these objects for a long time. In South Africa, I decided they were design responses to a technology that could not be purchased by poor people, whereas what I was teaching in the university derived from a German design pedagogy that eagerly embraced available modern technology. So I made a new version of my academic program and over a period of about six years taught it to local people at seven short-term summer schools in five African countries. We used anything that was available, often thrown-away rubbish. Passers-by dropped in off the streets and became students. Almost everyone responded to the exercises quickly and directly, often humorously. They seemed able to deal with issues of form with the same intensity and forthrightness of the boy in Durban.

Late one night I took some jazz musicians home to their black township on the southwestern side of Johannesburg. I had never been to Western Native Township before; whites did not go to such places. But I returned many times after to study the people and their houses, particularly the way they had plastered and painted the small boxes, which they had been renting from the municipality since the influenza epidemic of 1917. Over a few years a team of students and I documented the fronts of all 2,000 houses. The facades were patterns of rectangles, circles, and half-moons, a restricted palette of shapes from which a communal language had been assembled. So, instead of painting a hammer-and-sickle on his wall, the first local chairmen of the African National Congress chose an open circle with a scrrated edge from the community's menu of forms, which he then read as an industrial rotor hub, a symbol of Russian progress. A woman who ran an illegal Fah-Fee (a popular Chinese-based betting game) on loon painted her lucky symbol, a horse, on her wall but made the horse of common triangles and half-moons. From these bare houses with seven people per room came an astounding decorated urbanism.

No designer on his or her own could have invented the decorative language of the Western Native Township community, nor could any designer have chosen the personal example each house displayed on its facade. Designers have tried their hand at animating dull housing and produced only abstract stereotypes. But many designers have learned the difference between professional and popular knowledge. They no longer see buildings as disassociated from their context; they try hard to revel in environments of complexity and difference; they design permanent monuments badly and ephemeral events much better; they treasure the every-day in open societies; and they know when to invite others unlike themselves in and when to stand aside.

We will never know whether we have lost the naive genius of the little boy in Durban. We work in the hope that such ability will be available not only to those who are poor, excluded, and have to dream about the possessions of those a class above them. Some believe that new technologies may help us nourish the full universe of our abilities. We have yet to see this in action, especially for people for whom our technology remains chimerical. But, above all, we need a social environment in which we see the value of others and do not consign them and their objects to invisibility. And if this happens, we may not have to choose between Afrikaner steeples and Zulu radios.

Julian Beinart is a Professor of Architecture and a Director of the Joint Program in City Design and Development at MIT.

Question 3: What does your design make you think of?

I think of dignity.—M. Myaluza

It makes me think of a butterfly. I am fond of them.—P. Butelezi

I think of my brother-in-law who did it to signify his success in his divorce case.—Phillip Letatola

It makes me appreciate the beauty of art.—Rhoda Nkile

It reminds me of two things: cypress trees and the insignia of a diamond card.—Phoofolo

I think of the Queen's crown.
—Joyce Swartbooi

I think of the freedom of movement I had in WNT.—Johannes Maseke

It reminds me of the money I had spent on it.—Ruth

I think of wealth in the form of a diamond.—A. Mkhize

I think of nothing.—Joel Ngubane

I think of Chinese and Japanese flags.—S. Ramaphosa

It makes me think of tombstones and graveyards. It is a memorial now because WNT is dead.—Phiri

I think of a horse. I am a fah-fee woman; a horse is my lucky number.
—Martha Sidzatana

I think of a razor which together with the black colour signifies "danger."
—Ishmail Setlodi

It makes me think of my late mother.
—M. Malunga

It reminds me of my brother I have not seen for three years now.—Mashaba

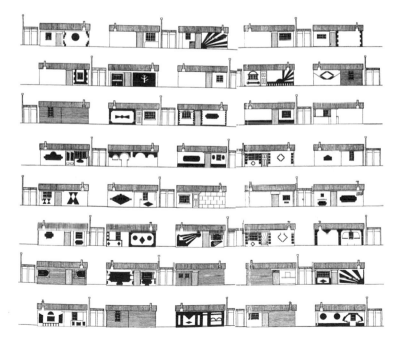

What imposes obligation in the present received and exchanged, is the fact that the thing received is not inactive. Even when it has been abandoned by the giver, it still possesses something of him. . . . In all this [giving and receiving] there is a succession of rights and duties to consume and reciprocate, corresponding to rights and duties to offer and accept. Yet this intricate mingling of symmetrical and contrary rights and duties ceases to appear contradictory if, above all, one grasps that mixture of spiritual ties between things that to some degree appertain to the soul, and individuals, and groups that to some extent treat one another as things.

All these institutions express one fact alone, one social system, one precise state of mind: everything—food, women, children, property, talismans, land, labour services, priestly functions, and ranks—is there for passing on, for balancing accounts. Everything passes to and fro as if there were a constant exchange of a spiritual matter, including things and men, between clans and individuals, distributed between social ranks, the sexes, and the generations.

—Marcel Mauss, *The Gift: The Form and Reason for Exchange in Archaic Societies*

THE BRACELET

Irene Castle McLaughlin

Silver, gold, shell, stone: my jewelry basket is organized according to these objective, formal properties, these visual markers and guides for ready access. Like any collection, my jewelry could just as easily be sorted by age, value, place of origin, or even by color. There is another underlying narrative that is known only to me. In the context of that story—my life story—these objects are heirlooms, gifts, invocations.

I reach for an old Navajo cuff bracelet when I want to invoke the spirits of my female ancestors and allies. The cuff is an object of power in its own right, a massive weight of heavy ingot silver that recalls medieval armor, or the wristband of a super hero. It was forged during the 1930s or 1940s, when southwestern silversmiths worked outside, or by the light of kerosene lamps, using only a few rudimentary tools.

During the formative years of Navajo silverwork, roughly 1880 to 1930, many men experimented with the alchemy of fire and metal. A certain self-reliant bravado was necessary for the elemental work of melting silver slugs over pinion coals, pounding out ingot bars with a hammer, and heating them with a home-made blowpipe or torch. When possible, silversmiths often made extravagant use of materials in those decades before Navajo jewelry became transformed into a form of fine art by innovators such as Kenneth Begay. My big bracelet is somewhat crudely made, but the design is unique and bold.

The imposing size and attitude of the bracelet is exaggerated by the fact that it was made to fit a tiny wrist. And as if to emphasize its inflexibility, it can be worn in only one orientation. The structure is an expanding spiral that doubles in width as it arcs from the inner to the outer wrist. The thick central band has been darkened to a smoky, matte black; the mark of a residual fire coat or its simulation. The contrast of dark and light enhances the exaggerated dimensionality of the design. Heavy triangular wires rim the sides of the band, which is laterally subdivided into four sections by lengths of wire punctuated at either end with big, globular, silver "raindrops." In the center of each section, huge turquoise stones are set into high silver bezels with irregular saw-toothed edges. Starting at the inner wrist, the size of the stones increases as the bracelet widens out. All of them are thick, domed, pieces of natural turquoise, left in free-form shapes. The first three stones look as though an unsteady hand splashed ethereal blue porcelain across the surface of rocks, leaving the grainy brown stones partially exposed.

These first three stones come from the historic Bisbee mine in Arizona, famous for having produced turquoise of great character. But it is the fourth and final stone, invisible to the wearer, that is the zinger, the masterstroke of the design. It is a huge, battered oblong of soft emerald green, delicately tinged with chartreuse and the blue of spring skies. This stone is a fabled variety known as Battle Mountain, a green phase of Blue Gem, which is considered by many to represent the apogee of American turquoise. And after the green orb, the finale: a row of six oversized "raindrops." There is no stamp identifying the maker; it is an anonymous monument to his mind and to the elements of fire, water, earth, and sky.

My bracelet is a weathered and venerable monument. Bracelets like this one were made primarily for an Indian audience and for Indian consumers, not for the tourists who disembarked from crowded railroad cars at the Grand Canyon and returned east with lightweight jewelry as signs of their travel. Among the Navajo, silver jewelry became a form of patrimony, a cultural aesthetic, an ethnic marker, an economic resource, a currency, and a standard of value. On the isolated reservations, both men and women wore profuse quantities of jewelry on special occasions—wagon trips to the trading post, dances and ceremonies, visits to relatives. Because they were creative acts, making and exchanging jewelry became deeply embedded in Navajo social life; it was believed that these practices helped to keep the universe in balance and in constant motion; they made things happen. Jewelry was loved and worn, given to relatives and friends, traded to Indians and non-Indians for coffee and sheep, turquoise stones and woven garments; converted to cash to pay for ceremonies and gasoline. My bracelet evokes that world of wood smoke, lambs, and matrilineal clans.

This bracelet, however, was destined to travel. At some point the Navajo who owned it pawned the bracelet at a trading post in exchange for cash or supplies and never redeemed it. At that point it briefly became a commodity in the high-end sector of the Indian arts market, in which pawn items enjoy a distinct cachet associated with their perceived authenticity. On a sunny Sunday during the early 1960s, a petite redhead named Irma Bailey purchased it at a trading post near Farmington, New Mexico, while driving with her husband Wayne. As she recalls, "He [the trader] had a collection of old pawn jewelry he was keeping in a train case or travel case. I dug around and bought several pieces out of that case and that was one of them."

Irma and Wayne were also traders, a lifestyle dictated by Irma's passion for Indian art and her affinity for Indian culture. As choreographed by Irma, Indian trading is as much a form of fictive kinship as it is a business. Like Polynesian chiefs, she is a genius at making relationships and keeping them in play, weaving them into an ever-expanding network of artists, clients, and friends bound by obligations and affections. This requires developing mutually interdependent relationships with native artists and nurturing them through a continuous series of exchanges: gifts and counter-gifts, debts and payments, trades of material, assistance and advice, purchases and commissions. It also requires cultivating a clientele of long-term customers and collectors who will fuel the system with cash, social capital, and other currencies. Irma is now in her ninetieth year. Conducted over the course of decades and across generations, these relations have become so layered, so seamless and embedded, that it is impossible for an outsider to differentiate between the social and the economic, to unravel the accounting.

Irma herself, however, knows just how everyone's ledgers are balanced; she has the eye of a connoisseur and a head for business. But it is her heart that drives the system, synthesizing it all into a seamless, fluid whole. She radiates a vitality that draws people into her orbit and compels them to respond in kind: few of her debts go unpaid. She conducts most of her business out of her Albuquerque home, which is alive with people from all walks of life. Here the artists have become her extended family, with all related responsibilities, joys, and sorrows. She feeds them at her table, shares Christmas with them, mediates disputes, buries their dead when necessary. In return, they bring her their best work and their friends, sometimes even name their daughters after her. Here, too, they gather other friends, some of whom might be called clients.

But again, this distinction rapidly dissolves in Irma's presence, which seems to implicate everyone in a larger system of meaning. Serious collectors are drawn here by the allure of rare pots, rugs, or jewelry. They buy things not only for their artistic value and cultural referents but also to capture an imprint of Irma's passage, her choices and experience. And so the balances are maintained, the cycle continues, the network grows.

Irma is seldom alone and rarely asleep. Her phone rings day and night; cars line the curb in front of her house. Indian ladies keep her cupboards filled with freshly baked bread, tamales, melons, and pies; customers send designer handbags, invitations, and introductions. Irma herself is often up cooking before dawn. Each day she starts the fire that fuels her universe.

Irma didn't sell the big pawn bracelet. It fit her diminutive wrist and she fell in love with its rugged charm, so she "wore it and wore it," through thirty-five years of Pueblo Feast days, a divorce and re-marriage, and as she traveled across the United States each spring to sell the things her artists brought her. A couple of years ago, she traded the bracelet to me in exchange for an equally imposing gold cuff that was made for Irene Castle, a famous dancer and stylist in the early twentieth century. Irene was my grandmother: I carry her name and inherited from her the fine bones that we share with Irma.

This bracelet exchange was not motivated by desire for fine jewelry; it was an expression of allegiance, a way of giving shape and substance to the intersection of three kindred women. My bracelet grounds me in an invisible social firmament, where Irene and Irma are stars in the constellations of descent and affinity. I feel their reassuring presence when the weight of the bracelet is on my wrist and I understand what it means to wear your wealth.

Irene Castle McLaughlin is an anthropologist and Associate Curator of North American Ethnography at the Peabody Museum of Archaeology and Ethnology, Harvard University.

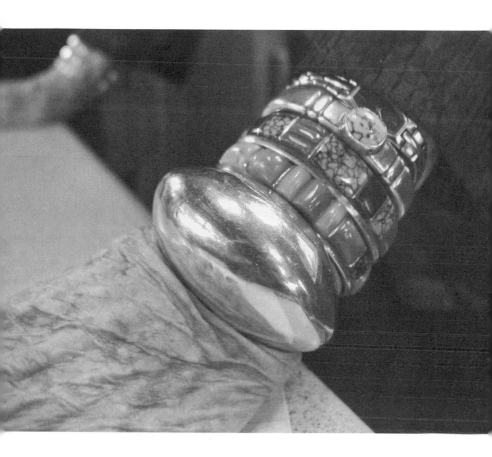

But how can we retrace our steps? Isn't the modern world marked by the arrow of time? Doesn't it consume the past? Doesn't it break definitively with the past? . . . Hasn't history already ended? By seeking to harbor quasi-objects at the same time as their Constitution, we are obliged to consider the temporal framework of the moderns. Since we refuse to pass "after" the [postmodern], we cannot propose to return to a nonmodern world that we have never left, without a modification in the passage of time itself.

—Bruno Latour, *We Have Never Been Modern*

THE AXE HEAD

David Mitten

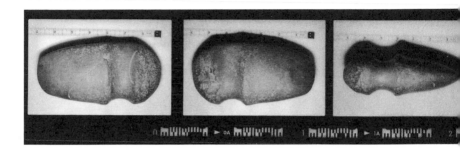

I first saw it at my maternal grandfather's house on his home farm, named Shannondale Farm, in Berlin Township, Holmes County, Ohio, more than sixty years ago. My personal talisman of the past, this stone axe head was made and used by native Americans around 5,000 years ago, in what was much later to become north-central Ohio. Its history stretches far back into remote time, thousands of years before European settlers came to this area of North America. Almost all my life, I have kept and treasured this axe head and the story it has to tell, one that can be read from its form, its surfaces, and the marks visible upon it.

Its material is a hard gray limestone that was originally part of a layer of calcareous sediment deposited on the bottom of a sea somewhere in what later became northern North America, many millions of years ago. Much later, these sediments were compacted, lifted, and squeezed over other millions of years until they hardened into a dense gray limestone. Still later, earth movements uplifted the layers of limestone onto dry land, where gradually a layer of limestone was exposed through erosion. Soon, chunks began to break off an outcrop of this limestone. One of these chunks was the mother of the axe head.

Many thousands of years ago, a huge, mile-thick continental ice sheet formed and grew until it covered much of North America. This massive glacier ground its way slowly southward, carrying with it millions of tons of soil and rocks, including this chunk of limestone. The end of the ice sheet stopped about sixty miles south of what is now Lake Erie. Churning streams of water, running under the ice sheet as it began to melt and retreat, rolled our chunk of limestone over and over countless times, abrading its surface, rounding off its jagged edges and corners, and finally smoothing it into what must have looked like an elongated, slightly flattened oval cobble.

As the glacier retreated north, it left rich moraines, deposits of earth, gravel, cobbles, and boulders on a gently rolling landscape. Vegetation, first tundra-like grasses, then bushes and dwarf trees, and finally dense deciduous forests of oak, maple, ash, and tulip poplar grew on the hills and valleys, as rivers and their tributary streams cut through the glacial deposits. One of these streams probably exposed our oval, rounded rock. How long it lay on the surface or in a stream bed, we have no way of knowing.

Now people had come into this country. They followed the first pioneering hunters who had crossed from Asia onto the North American land mass at least some 12,000 to 15,000 years ago, pursuing the big mammals— mastodons, bison, horses—that were their principal prey. The later native people hunted, fished, and gathered seasonal fruits, nuts, and berries. They settled camps and villages in the river bottoms, where they grew crops: maize, squash, and beans. For their tools and weapons, they constantly sought suitable flint and chert from the rich outcrops in this area, for flaking into spear and arrow points, knives and scrapers. They also looked for hard, dense stones included in the moraine deposits, stones that had been carried by the ice sheet from hundreds of miles to the north: granites, quartzes, volcanic rocks, and limestones.

A man sorting through these stones finds our cobble. He was searching for a stone suitable for fashioning into an axe head that would help him fell trees, split logs, and dig charcoal out of the timbers that would become dugout canoes. Our cobble fit the bill perfectly. He picked it up and took it back to his camp. Then he (and perhaps others, taking turns) began the slow, laborious job of imposing functional form on the stone. He must have already had a visual image of what the finished axe head would look like, as well as other similar axe heads that his people had already made and used. Seemingly endless pecking, pounding, chipping and grinding took place, probably over several years, much of it probably during the winters, when there was enough leisure to devote to such time-consuming work.

My father, a teacher in the Cuyahoga Falls High School, once assigned a project to his students: to make a ground stone axe. The students thought that this would be easy, but after a week they had barely pecked or ground a single mark on the granite cobbles that they had been given. A major part of this sculpturing process was pecking, abrading, and polishing a deep, rounded groove three-quarters of the way around the axe head. At the same time, the blade of the axe head was pecked and ground down, so as to leave a slight raised edge on the side of the groove toward the blade. One-quarter of the axe head was smoothed along its entire length. Further grinding and polishing added a subdued luster to the curving edge of the axe head and the two adjacent surfaces. It didn't matter if large areas of pecking marks were left on the flat edge and on the oval, rounded butt, opposite the cutting edge.

Finally, the maker was satisfied. He had chipped and pecked and ground away much of the cobble that he had found, transforming it, freeing it (almost in the sense that Michelangelo so many centuries later longed to free the forms hidden, imprisoned in his blocks of marble), into a well-crafted object, highly suitable for cutting and chopping and a work of beauty as well.

In the maker's time, function and aesthetic sensibility were inseparable. In our own time, we have come to appreciate keenly how much our aesthetic appreciation and understanding of an object, or a work of art, can be enhanced by learning how it was made, how the mind, eye, and hand of the creator had imposed his desired form on the raw material.[1]

The maker fashioned a stout handle of some tough yet flexible wood, perhaps from the branch of an ash tree. He split one end, which he softened in water and gradually bent around the axe head, fitting the two halves into the groove. He then tied the split ends with wet rawhide thongs and wrapped these thongs around the shaft. He also inserted a wooden wedge, perhaps also of ash or oak, under the rawhide bindings along the flat side of the axe head. When the rawhide dried, it shrank, binding itself firmly onto the stone in a solid grip that would withstand the constant, hard blows to which the axe head would be subjected. Finally, the maker may have performed some ceremony over the finished axe furnished with its handle, perhaps to insure that it would cut well and last for many years.

The marks on the surface of the axe head are mute reminders of countless blows that were inflicted on it—chopping down trees, splitting logs into firewood, hollowing out tree trunks into dugout canoes. On one side of the butt are breaks where spalls of the limestone broke off and fell away; that the axe continued to be used is shown by the smoothing on these scars. Beveling on the cutting edge suggests at least one episode of resharpening. Although a diagonal crack appeared curving across one face of the axe head, it did not lead to a major break. At the same time, the appearance of this crack may have led to the discarding of the axe head, or at least to using it for less strenuous cutting tasks.

The surface of the axe head is covered with tiny scratches and incised lines. Examination under a microscope would surely yield much information about how this axe head was used and against what kinds of materials. One mark remains mysterious: a pecked or incised line within the binding groove. Was this added by another owner, who began to alter the shape of the axe head, or could it have been damage inflicted by a nineteenth- or early twentieth-century plowshare?

The span of use of this axe head could have covered decades or perhaps several generations of these people's lives. It may have been handed down from father to son, or nephew, or even to a grandson. It may have been used by women as well, and indeed it could well have been a woman who made it in the first place. A treasured possession, it may also have signified rank, status, or prestige. We cannot know.

Somehow, at some time, the axe head was discarded or lost. For many centuries, it remained on or under the soil of the hill slopes leading down to Martin's Creek while, across the world, Mesopotamians built cities and alphabets, and, in the Nile Valley, the great pyramids were rising. One day, early in the twentieth century, maybe eighty to one hundred years ago, a man found it.

Now the Europeans had come and had settled here, cutting down the forests and turning much of the land into pastures and cultivated fields. One part of this land had become a farm, owned by the Boyd family. The owner, Umfrey H. Boyd, found the axe head, perhaps while plowing behind his team of horses, perhaps just while walking the bounds of his land. He recognized it as something that had been fashioned by human hands. Picking it up and wiping off the brown soil that covered it, he brought it back to his farmhouse and washed it. Probably his wife and the children—three daughters and a son—admired it, and perhaps begged their father to tell them a story about the axe head and Indians who had used and lost it. Then it was placed in a corrugated cardboard carton to join other "Indian relics": arrowheads, spear points, scrapers, and parts of two tubular tobacco pipes made out of gray and black slate.

One of the daughters, my mother, Helen Louise Boyd, married a tall, handsome schoolteacher from the other side of the county, Joe Atlee Mitten. In due course, as a little boy, I visited my grandfather's farmhouse and remember spending many hours playing with the stone objects collected in the carton, kept in a cool, fragrant back room. I especially remember holding the axe head, feeling its smooth weight, its polished, pecked surfaces, and musing on the long-ago, long-dead people who had walked these same hills and creeks and had left such an evocative, silent witness to their vanished presence. I wondered about all their patient, persistent effort that had succeeded in shaping the beautiful and functional abstract form that they desired upon such dense and intractable stone.

I owe it a great deal, for it serves as a special talisman, perhaps even a guide, that led me to my lifelong path as a teacher and archaeologist of those ancient cities and alphabets and peoples of the Mediterranean world. The axe head still rests atop the mantel of my fireplace and will until I too recede, no longer telling but told into its long story by my daughters, who will I hope remember its deeper provenance in their own way, in their own time.

David Mitten, whose lifelong research is on classical bronze statues and vessels, is James Loeb Professor of Classical Art and Archaeology at Harvard University and George M. A. Hanfmann Curator of Ancient and Byzantine Art, Emeritus, of the Harvard University Art Museums.

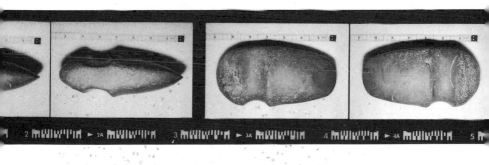

The whole universe of concrete objects, as we know them, swims . . . for all of us, in a wider and higher universe of abstract ideas, that lend it its significance. As time, space, and the ether soak through all things . . . good, strong, significant, and just. Such ideas, and others equally abstract, form the background for all our facts, the fountain-head of all the possibilities we conceive of. They give its "nature," as we call it, to every special thing. Everything we know is "what" it is by sharing in the nature of one of these abstractions. We can never look directly at them, for they are bodiless and featureless and footless, but we grasp all other things by their means, and in handling the real world we should be stricken with helplessness in just so far forth as we might lose these mental objects, these adjectives and adverbs and predicates and heads of classification and conception.

This absolute determinability of our mind by abstractions is one of the cardinal facts in our human constitution. Polarizing and magnetizing us as they do, we turn towards them and from them, we seek them, hold them, hate them, bless them, just as if they were so many concrete beings. And beings they are, beings as real in the realm which they inhabit as the changing things of sense are in the realm of space.

—William James, *The Varieties of Religious Experience*

DIT DA JOW: BRUISE WINE

Susan Spilecki

On my desk stand two small plastic bottles. On the label of the first are the words *Dit Da Jow,* Wing Lam Secret Kung Fu Liniment. Wing Lam is a teacher of kung fu in Sunnyvale, California, by way of Canton and Hong Kong. The bottle label sports an antique photo of an older Chinese man, shirtless, beating his open palm against a stack of bricks. Behind this photo is the outlined picture of a tiger fighting a dragon. Twisting off the white cap, I can smell the liniment (*dit da*—bruise or strike; *jow*—wine or tincture). It smells like a mix of soy sauce, unrecognizable herbs, and a little alcohol. It looks brown, but if I pour a few drops into my palm and rub them over my sore wrist, they do not stain my skin.

The liquid in the other bottle is darker and more pungent, the scent more like a mix of ninety-proof whisky and myrrh. The bottle is also larger, but it has no Chinese characters. In fact, the bright green label claims that the bottle contains Poland Spring Natural Spring Water. This is a lie. It contains my teacher's *dit da jow*. The recipe is the one he got from his teacher, Tang Kwok Wah, who got it decades ago from his teacher, Lam Sai Wing, who got it from his teacher, Wong Fei Hung. If Americans have heard of any of these men, it would be Wong Fei Hung, about whom Jet Li and Jackie Chan have made several movies, portraying the greatest of the martial arts champions of southern China as a kung fu master of the old style, a man who could both kill and heal. Those killing and healing arts came down from the Shaolin Temple, from the seventeenth century and further back in time, before the Ching Dynasty burned it down and killed all the kung fu masters but the five who managed to flee.

"The kung fu you can see isn't the real kung fu," my teacher says as we sit in the coffee shop after class one evening. "Those karate guys with their big muscles and stiff punches? That's all crap."

On my left sits Sean, a burly Boston Irishman who studied White Crane kung fu before he came to Sifu. On my right sits Andy, a laconic man in glasses, given to questioning everything Sifu says. Sifu, my teacher, is an unprepossessing Chinese man of average height and build, with grey hair, although he is only in his early fifties. You'd never think he could send a much bigger, muscle-bound fighter across a room with a single punch, although his students know that he has done it more than once.

He holds out his hands for our inspection. There are no calluses. He grins.

"Smooth as silk. When I'm eighty, I'll still be able to kill those guys, but they'll have arthritis. Because they only saw part of the technique. They didn't see the *jow.*"

Looking up *dit da jow* on the Internet, I find that there are three main types of herbs used in making it: anti-rheumatic herbs, herbs that stop bleeding, and blood-circulating herbs. The anti-rheumatic herbs have "pain-relieving (analgesic), anti-inflammatory, and circulation-promoting" properties.[1] In the simplest terms, getting more oxygen into the blood, and getting the blood circulating more freely around injuries, speeds healing. Sifu, selling his *jow* only to us, his students, can keep his recipe secret. The kung fu master Wing Lam who sells his product on the Web, doesn't have that option; the law requires him to list his fifteen ingredients in decreasing order of importance. Even so, he manages to make it difficult for those who might steal his secret: he lists the herbs in Latin. I look them all up on the Internet and can decipher only six: ginseng, caesalpinia sappan, safflower, mastic, myrrh, and dragon's blood.

These are respectively an anti-inflammatory, an astringent, a diaphoretic, a stimulant, an antiseptic, and another astringent. Whatever you call them, when they're put together properly, they work.

It is my second class with Sifu and the guys. I am punching Andy barehanded. His chest protector resists, and I end up with a blood blister between my fingers. I shake out my hand to relieve the pain.

Sifu says, "Who's got *jow?*"

Patrick, our immunologist, pulls out a Poland Spring bottle filled with brown liquid. Sifu takes it and applies the liquid to my hand, rubbing it in thoroughly. The next day, I can see no sign of the injury.

I bruise easily. I always have. So from the beginning, I am impressed, if surprised by the *jow*. My mind opens a crack. Weeks pass. Every Wednesday and Friday evening, under Sifu's supervision, we practice forms, attack each other, and point out flaws in each others' defenses. Occasionally, I get injured from the normal connection of arms and legs that martial arts study entails—a bruise here, stiffness there. The *jow* speeds my healing, and I am beginning to see the light.

The old Chinese masters had two kinds of students: outer-door and inner-door. The outer-door students were like us, regular students who learned the basics and went home. The inner-door students each got a little time behind closed doors with the master. He gave each of them a bit extra, but he didn't give any of them a whole lot, not enough to mess with him, anyway.

The issues of secrecy and caution pervade the history of kung fu. The Ching government justified burning down the Shaolin Temple in 1722 by claiming that the Buddhist fighting monks who lived there were Ming sympathizers, working to build a rebel army to restore the Ming emperor and kick out the Mandarin Ching rulers. If this was not true before the Ching burned down the temple, it certainly was afterward. The surviving remnant, tradition says, began the Hung Society, *hung* being a word that could refer to the Ming Dynasty's founder and to the *hung soan,* the red houseboats that were used as homes, transportation, and training space for the opera troupes that traveled the waterways of Canton Province. Hiding among the opera troupes, which routinely used martial arts and acrobatics in their traveling shows, the Hung rebels could keep moving, proselytize, and train other potential rebels. And they practiced the deadly arts that would (they hoped) guarantee their vengeance. The style was well suited to practice on boats, with its compact stances and short-range hand techniques. Tradition claims that the years afloat modified it further. To this day, Hung Gar kung fu relies on few kicks compared to other martial arts, such as Japanese karate, Korean tae kwon do, Indonesian pensilat, Thai kickboxing, and even kung fu styles from northern China.

The Ching Dynasty continued until 1911. The next few decades of nationalism were not a golden age for kung fu practitioners, but they were better than when Mao rose to power in the 1950s. He hunted down kung fu masters and killed them as a matter of policy, knowing that they were men who thought for themselves and could, with their bare hands, kill anyone who didn't like it.

Thus began the second kung fu diaspora. The surviving masters fled to Hong Kong, Taiwan, and the Philippines. A few went further, to Europe and North America, but they remained understandably cautious.

This long tradition of secrecy helps explain my teacher's disdain for just about every other fighting style on earth except his teacher's. For Sifu, any style that didn't come down from his teachers, and cannot show a direct connection in its lineage with the Shaolin Temple, has no validity.

"The old masters," Sifu says, "You think they teach this out to just anybody?" He shakes his head at our ignorance. "It's not about the money. They don't care about money! It's about danger. If I'm an old master and

I teach you my technique, you can kill me. So I only teach a little bit. A bit to you, another bit to you, and you." He points to us each in turn. "So I'm safe."

"Nobody has the whole thing?" asks Sean.

"Exactly! Once they teach it out, they can't take it back. So they're careful what they teach. And who they teach."

"But you teach us all the same," I say.

He shrugs. "Things are different now. In the old days, students lived with the teacher. They did kung fu twenty-four hours a day. Nowadays, nobody's got time. You guys do punching practice for an hour a day at best." He rolls his eyes again.

He is right of course. He is in no danger from us. And while, on the one hand, I don't think I'd want to eat, sleep, and breathe kung fu day and night for the rest of my life, I do understand what I'm losing—what we are all losing, Sifu included—by living "normal" twenty-first century lives. We each work forty to fifty hours a week: driving a bus or a train; teaching Latin to high school kids, writing and literature to college kids, exercise to the elderly; answering customer service calls; researching AIDS. We spend an hour or two a day stretching, doing forms, punching the bag, practicing with the sword and/or staff, maybe some sparring, maybe not. It's never enough, not if what we're after is mastery of combat techniques.

But I'm not sure that's what we're after.

For us, I think it's more about getting exercise, participating in history, having friends—a family—who also care about this obscure, dying art, this long, proud lineage. But that is probably why our teacher gets so frustrated with us.

Back in late November, I hurt my wrist, probably not even in kung fu class. Every night, I rubbed Wing Lam's *jow* on it, because I had already worked my way through half of that bottle and didn't want to start the new bottle I had bought from Sifu until I had finished the other. But my wrist got no better, so finally I decided that I had to ask Sifu for help. I considered bringing Wing Lam's *jow,* but at the last minute packed Sifu's unopened bottle instead. Thank heavens.

"Sifu, my wrist still isn't better."

"Let's see."

I held out my hand and he manipulated it, felt for swelling, dislocations and God knows what else.

"It feels all right," he said. "You probably just pulled something. You have *jow?*"

"Right here." I handed him the bottle.

"What's this? You haven't even opened it!"

"I was using the bottle I got on the Web—"

"You say, 'Oh, Sifu, my wrist hurts!' and 'Oh, Sifu, can you fix it?' but you don't do what I tell you!"

"But I thought I should—"

"I don't want you to think! You have to use my *jow.* It looks like the others, but it's different. It's like our kung fu: it looks the same, but it's not!"

He rubbed his *jow* into my wrist fiercely, and I tried hard to look penitent and contain my winces. I wanted to say, "Sifu, my father grew up during the Depression. I'm genetically unable to waste anything." But I knew it wouldn't do any good. Instead, I promised to use his *jow.*

Is it my imagination that Sifu's *jow* works better than Wing Lam's? Certainly Sifu's *jow* is fresher, coming to us within days of his making it. And it definitely smells stronger. So maybe it's the value of the personal interaction, which is also part of the tradition. Wong Fei Hung's students got their *jow* from him, not from some teacher in Beijing who packed it on a donkey and sent it south. Maybe it's just a better recipe. Or maybe it's the strange love and fierceness Sifu seems to have for us, his hapless, modern American students, that gets transmitted into the chemistry.

Whatever. I use Sifu's *jow* because Sifu tells me to. And in the end, that's part of the tradition, too.

Susan Spilecki teaches writing and literature at Emerson College and Northeastern University.

A commodity appears at first sight an extremely obvious, trivial thing. But its analysis brings out that it is a very strange thing, abounding in metaphysical subtleties and theological niceties. So far as it is a use-value, there is nothing mysterious about it, whether we consider it from the point of view that by its properties it satisfies human needs, or that it first takes on these properties as the product of human labour. It is absolutely clear that, by his activity, man changes the forms of the materials of nature in such a way as to make them useful to him. The form of wood, for instance, is altered if a table is made out of it. Nevertheless the table continues to be wood, an ordinary, sensuous thing. But as soon as it emerges as a commodity, it changes into a thing which transcends sensuousness. It not only stands with its feet on the ground, but in relation to all other commodities, it stands on its head, and evolves out of its wooden brain grotesque ideas, far more wonderful than if it were to begin dancing of its own free will.

—Karl Marx, *Capital: A Critique of Political Economy*

THE VACUUM CLEANER

Nathan Greenslit

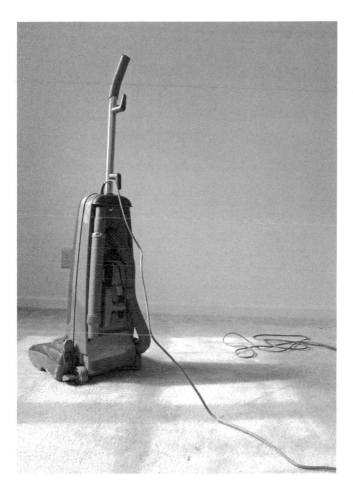

When she was about two years old, my daughter Emma was afraid of our rather loud vacuum cleaner. When children get scared, adults try to comfort them by representing their fears as irrational. Adults say: "It's okay, but see, there's nothing to be scared of." In this negotiation, children don't buy this adult story for a second; they come up with their own plans to manage their feelings. In Emma's case, she talked to our vacuum, telling it that "it's okay that you scare me." Emma's strategy was to help the vacuum deny its own identity as a frightening object. Indeed, part of what was so scary to Emma about the vacuum's loudness was that it evoked not so much the question of what this thing was, but rather who this thing might be. For Emma, the vacuum was something that needed to be dealt with like a person. And as a parent, I sometimes felt like a reluctant Sancho Panza to my child's quixotic adventures in self-discovery.

Once the belt drive broke in the vacuum, and I showed Emma how it worked. But when I asked Emma if she wanted to help me replace the belt, she refused. She did not want to touch the insides of the vacuum, as if it might start up without warning, as if it were just pretending to be broken. (Like when I would pretend to be sleeping and tickle her when she tried to sneak by.) She didn't trust that the vacuum couldn't animate itself without electricity from the wall. Children teach us everyday forms of what philosophers discuss in seminar rooms: Aristotle devised a system of motions that characterized self-locomotion as belonging to the category of the highest forms. But I couldn't convince Emma that the vacuum was not this kind of animal. Emma's concern that the vacuum might spring to life combined her Aristotelian intuitions and an element of the original concept of stress—that it is not harm per se, but rather the unpredictableness of a potentially harmful environment that strains the nervous system. Indeed, the term *stress* was introduced in animal research during the 1950s, research that demonstrated the effects of randomly shocking monkeys. The random shocks produced greater effects than shocking the monkeys with regularity.[1]

Our popular use of the word *stress* is relatively recent. Stress as a scientific term was introduced in the 1950s to name the bodily effects of new demands: a frenetic and socially competitive lifestyle and a household rich in ambient noise. In scientific circles, stress was commonly described in terms of modern anxieties.

Thus, the scientific understanding of the nervous system provided social commentary on industrialization. Biology was not seen as timeless, but as a way to understand our relationship to a particular social and cultural world. Here, the vacuum cleaner is an actor—loud and stressful. In *Capital*, Karl Marx makes it clear that when objects become commodities they become nearly animate.[2] The French philosopher and sociologist Jean Baudrillard takes the activity of consumer objects and puts them in a larger context. Taken individually, consumer objects have no meaning—that comes from their participation in a *system* of objects.[3] In this framework we don't consume individual objects; we consume the social order that they belong to. We buy the vacuum; we consume assumptions about gender, households, families, and social status.

In response to Emma's fears, my wife, Heather, asked Emma to help her decorate the vacuum with butterfly stickers and encouraged Emma to dress up the vacuum in goofy outfits, including long scarves and an old fedora hat. Also, Heather would give Emma some of our vacuum's cleaning attachments to play with. I remember one time Emma walked around with the extension hose draped over her shoulders as a stole, with another attachment as a cane, and said she was "borrowing" these accessories from the vacuum. For her, the vacuum "had" parts—it possessed parts—and could share them with a playmate.

About six months later, Emma's behavior toward the vacuum cleaner changed. She began leading her younger sister, Ellie, into a frenetic dance whenever we vacuumed the floors. Ellie herself was never scared of the vacuum, but now her sister was teaching her how to not be afraid of it all the same. To this day, when Ellie is by herself, she seems indifferent to the vacuum, but when Emma is around, Ellie lets her sister minister to her as Emma leads her sister in a ritual of laughter, shouts, and dance.

When Emma began this dancing rite, I asked her why. She answered, "Because we're dancing, Dad!" Exactly. This is what two-year-olds do. The explanation they provide for some action is the action itself. They don't rationalize their experiences, they narrate them as we adults end up teaching children that behaviors have meanings and motives and that we might not know why we behave the way we do. When Emma got scared of the vacuum, she never asked, "What's wrong with me?" She wondered, "What's wrong with this thing?"

Toy vacuum cleaners are friendly, colorful, and make "fun" sounds. They are marketed as ways to help children deal with their fears of "real" vacuum cleaners. ("See, vacuum cleaners aren't really that scary, are they?") The toy makers say to us: This is what your children want—this is what they want vacuum cleaners to be. Once I asked Emma if she wanted one of these toy vacuums. She refused when she found out it wasn't a "real" vacuum, that it was "only" a toy. Of course, wasn't it the reality of our vacuum that scared Emma so? But here was this new object, designed like a real vacuum minus the terror, and Emma wanted nothing to do with it. She didn't want objects to be figured out for her. She wanted to be able to figure objects out for herself.

I'm interested in those moments when we momentarily see the child within us. When grownups are startled by something—a machine turns on without warning—we react by rationalizing, "Oh, it was just the blender," or "Phew, it was just a car." For a moment, we have experienced the world as truly animate and have a genuine connection with our usually invisible histories with objects.

Our psyche reveals its structure when the everyday wobbles, when what we've always taken for granted slips out of place and suddenly appears strange. For me, this has been the most profound part of being a parent. There are moments when I see myself in my children. At one point, the crazy Don Quixote turns to Sancho Panza and says, to paraphrase, "Consider what you are and try to know yourself, which is the most difficult study in the world."

Slowly, Emma is becoming less interested in the vacuum. Now, it largely annoys her. For instance, she might be watching *Blue's Clues* and ask us to wait until the program is over to finish vacuuming the living room. Emma is now three and a half. A couple of days ago, I asked her about the vacuum cleaner. "What do you think about the vacuum? Do you remember when it used to scare you?" Emma said, "Yeah. It's still loud. I like quiet things better. I think everybody likes quiet things better." For me, that vacuum cleaner is all the more evocative precisely because it has ceased to be special to Emma. It has disappeared into her psyche as just another object in a mundane world that she can start to take for granted. Emma is growing up.

I have a different relationship with our vacuum after having watched Emma figure it out for herself. Its noises mean something different to me. I wonder who is around to listen to them. Of course, I'm around to listen to them. And I find myself wondering who I am, such that my own psychic history with the vacuum—the time when it was animate—has gone invisible. So, it turns out that the real evocative object here is my daughter Emma, in whom I see myself, despite myself. This is one version of Freud's uncanny—things known of old yet somehow unfamiliar—that we as persons, who spend a lifetime trying to fantasize our way out of being children, can't help but see ourselves in our own children. We say it is cute that children have "unreal" relationships with objects. But just past this, is what's evocative: by attending to the experiences of children we gain much insight into why we so need to categorize the world. Every great once in a while, the vacuum cleaner startles me too.

Nathan Greenslit earned his doctorate in the Program in Science, Technology, and Society at MIT.

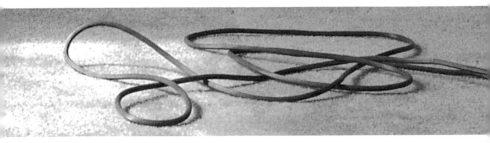

Objects of Transition
and Passage

Dreaming makes *everything in me which is not strange, foreign,* speak: the dream is an uncivil anecdote made up of very civilized sentiments. . . .

What is significance? It is meaning, *insofar as it is sensually produced.* . . . Then perhaps the subject returns, not as illusion but as *fiction.* A certain pleasure is derived from a way of imagining oneself as *individual,* of inventing a final, rarest fiction: the fictive identity. . . . I write myself as a subject at present out of place, arriving too soon or too late. . . .

If it were possible to imagine an aesthetic of textual pleasure, it would have to include *writing aloud. Writing aloud* is carried . . . by the *grain of the voice,* which is an erotic mixture of timbre and language and can therefore also be, along with diction, the substance of an art. . . . Its aim is not the clarity of messages. . . . What it searches for are the pulsional incidents, the language lined with flesh, a text where we can hear the grain of the throat, the patina of consonants, the voluptuousness of vowels, a whole carnal stereophony.

—Roland Barthes, *The Pleasure of the Text*

THE MELBOURNE TRAIN

William J. Mitchell

I was born in a lonely flyspeck on the absurdly empty map of the Australian interior. When I eventually took an interest in such things, I discovered that Mark Twain had once passed through there, and had written in *Following the Equator:* "Horsham sits in a plain which is as level as the floor—one of those famous dead levels which Australian books describe so often; gray, bare, somber, melancholy, baked, cracked, in the tedious long droughts, but a horizonless ocean of vivid green grass the day after a rain. A country town, peaceful, reposeful, inviting, full of snug houses, with garden plots, and plenty of shrubbery and flowers."[1]

We moved away when I was very small, but I still remember the river—arched over with red gums, and loud with the sound of magpies, kookaburras, and the occasional screech of a cockatoo. You could stand on the bridge and drop stones to plonk into the muddy water. There was a broad main street, with shop verandas and angle parking for the few cars. The baker, the milkman, and the iceman delivered from horse-drawn carts. Across the Natimuk Road were dry, grassy paddocks, and my dad always carried a big stick for the snakes when we walked there. Old Baldy Anderson (though nobody called him that to his face) ran the pub.

Every evening, the express train from Melbourne came thundering into town—passing through, and barely pausing, on its way to Adelaide. You could hear the whistle blowing—with urgently increasing intensity, then a mournful, gorgeous Doppler shift—from miles away across the starlit plains. The locomotive was a magnificent smoking, hissing, clacking monster sporting a glowing firebox, a tender heaped with filthy coal, and huge, shiny wheels. It was my earliest intimation of the technological sublime.

Throughout my bush childhood, the trains served as mobile metonyms for a wider world. In the slang of the day, the sprawling coastal cities were "the big smoke," and the steam engines were the fleeting local bearers of that emblematic attribute. They puffed great clouds of it up into the otherwise perfect hemisphere of clear blue sky, and left long plumes trailing across the flat horizon—matched, occasionally, by the dust plumes from cars speeding along dirt roads. When you entered a tunnel on the train, you had to leap up to close all the windows; otherwise, your compartment filled instantly with choking soot.

Each warmly lit carriage interior was a synecdoche of urbanity—an encapsulated, displaced fragment of the mysterious life that was lived at the end of the line. The passengers dressed differently from the locals, and they talked of unfamiliar things. They carried with them the Melbourne newspapers—the sober and serious broadsheet the *Age,* the racy *Sun* and *Argus,* the evening *Herald,* the *Sporting Globe* (printed, for some reason, on pink paper), and the utterly scandalous tabloid *Truth.* News was scarce in the bush, in the days before portable radios and casual long-distance calls, so fresh papers were eagerly awaited; passengers would sometimes toss them out to the railway workers who stood leaning on their shovels as a train groaned slowly by, much as they might offer a smoke to a stranger, or slip some flour and tea to a swagman at the door.

The passenger compartments were beautifully crafted in polished wood, overstuffed leather, screwed brass and chrome fittings, frosted glass with railway insignia, heavy sliding doors that closed with a satisfying thump, and little enamel notices enumerating prohibitions—spitting, smoking in the wrong places, frivolously pulling the emergency brake chain, and flushing the toilet while the train was stopped at a station. They were meticulously equipped with hooks for the broad-brimmed hats that all the men wore, ashtrays for the heaped remnants of cigarettes (some old-timers, I observed with amazement, could casually roll their next smokes with one hand while stubbing out the last with the other), overhead racks for suitcases, and chemical foot warmers that you would take out from under the seats and shake to activate. And there were wondrous cabinets of curiosities, with friezes of large, sepia photographs over the seats—each one depicting a ferny gully, a gravel track lined by huge eucalyptus trees, a mountain lookout, a wild patch of coast, or some other picturesque scene from the extensive territory served by the Victorian Railways. When I was a little older, and my family had picked up and moved to the shores of the Southern Ocean, the Jubilee Train came to town—a celebration of the fiftieth anniversary of the federation of the former colonies and formation of the Australian nation. The Jubilee train overflowed with the vast, varied, and unruly world distilled into a collection of mementos and souvenirs. I saw famous gold nuggets, the bullet-dented armor of the outlaw Ned Kelly, creepy remnants of the cruel convict era, stuffed birds and animals, diving helmets, feathery coral, miscellaneous minerals, and giant clams from faraway Queensland.

It was on a train, long before I was reluctantly dragged off to school, that I first realized I could read. With my nose up against the window, I began to decipher the signs advertising Bushell's Tea, the mileage markers that crept by, and the names of the stations where we creaked to successive stops—words in memorable sequence, the beginnings of narrative. I quickly found that the made-up narratives of books vanquished the boring hours as we crept across the plains. It wasn't long before I ran through the meager supply of kids' books, and moved on to the volumes of Henry Lawson that I had discovered at home. Lawson, to my gratified astonishment, wrote not of the Old Country and the Empire, nor of exotic American adventure, but of the people and places I *knew*. He was the bard of the bush. I loved the deadpan desolation of his great stories "The Drover's Wife" and "The Union Buries Its Dead." I could readily have believed that his famous character Mitchell the bushman, arriving with battered swag and old cattle-dog at Sydney's Redfern railway station, was a long-lost uncle. I was stirred by his angry anthem of the under-dog, "Faces in the Street." And sometimes it seemed that he was sitting beside me, gazing out into the shimmer-ing distance:

> By homestead, hut, and shearing-shed,
> By railroad, coach, and track—
> By lonely graves of our brave dead,
> Up-Country, and Out-Back:
> To where 'neath glorious clustered stars
> The dreamy plains expand—
> My home lies wide, a thousand miles
> In the Never-Never Land.[2]

It didn't matter that he had some patch of Western Queensland in mind when he wrote those lines. It didn't matter that he had died, drunk and penniless on the streets of Sydney, decades ago. I knew exactly what he meant. The power of his words, magically locking on to the landscape before me, made him vividly present.

When I was learning to write schoolboy essays of my own, perched at a wooden desk with porcelain ink-well and steel-nibbed pen, I often thought of sentences as trains. You could shunt the words around, like rolling stock on a siding, until you got them in exactly the right order. Like empty boxcars, they could carry the freight of simile and metaphor. And verbs, surely, were loco-motives. Put them up front for snappy imperatives. Mul-tiply, mass, and combine them for extra power. Keep it short. On the other hand, if the mood took you, and you wanted to construct a long, slow, freight-train of a sentence, with reflective asides in the manner of writers like Joseph Furphy, you could just let a few scattered verbs help it along from somewhere in the middle. Or, for a different effect, they might follow, pushing. When I memorized and recited poetry from the *School Reader*—mostly jingling ballads, like "The Wreck of the Hesperus" and "The Man from Snowy River"—the rhythms of the rails were always on my mind. Eventually, I got to read Pope on poetry, and realized he was right: the sound must seem an echo to the sense.[3]

As the years went by, and I made myself into an architect and urbanist, I began to understand that objects, narratives, memories, and space are woven into a complex, expanding web—each fragment of which gives meaning to all the others. For me, it was a web that grew from a quiet, isolated place on the banks of the Wimmera River.

It is more than half a century, now, since I left that little town. A decade after leaving, when I had the chance to attend Melbourne University, I fled the bush forever and have since lived my life among the world's great cities. But the sight of an express train still evokes the other end of the line. Now it recovers the memory of a spreading, aromatic peppercorn tree, a corrugated iron roof that was too hot to touch when you climbed up to retrieve a ball, the sudden smell of raindrops in the dust, and a small, curious child—walking with his impossibly young and beautiful parents along a silent, sunburned street.

William J. Mitchell is Alexander W. Dreyfoos, Jr., Professor of Architecture and Media Arts and Sciences at MIT.

In doing the biography of a thing, one would ask questions similar to those one asks about people: . . . Where does the thing come from and who made it? What has been its career so far, and what do people consider to be an ideal career for such things? What are the recognized "ages" or periods in the thing's "life," and what are the cultural markers for them? How does the thing's use change with its age and what happens to it when it reaches the end of its usefulness? . . .

The biography of a car, [for instance], would reveal a wealth of cultural data: where it was acquired, how and from whom the money was assembled to pay for it, the relationship of the seller to the buyer, the uses to which the car is regularly put, the identity of its most frequent passengers, and of those who borrow it, the frequency of borrowing, the garages to which it is taken and the owner's relation to the mechanics, the movement of the car from hand to hand over the years, and in the end, when the car finally collapses, the final disposition of its remains. All of these details would reveal an entirely different biography from that of [any other] car . . . . We accept that every person has many biographies . . . each of which selects some aspect of the life history and discards others. Biographies of things cannot but be similarly partial.

—Igor Kopytoff, "The Cultural Biography of Things: Commoditization as Process"

1964 FORD FALCON

Judith Donath

In 1964, when I was two, my mother bought a baby blue Ford Falcon. She drove this car for several years while I sat in the back, asking if we were there yet. Then we moved to Long Island, home of the endless traffic jam. The Falcon's response to the slow stop-and-go crawl was to overheat, its red emergency light insisting on a stop by the side of the road, hood open, radiator steaming. My mother bought a more reliable car and my father decided to store the Falcon in the garage, in anticipation of the day when he would obtain his American driver's license. Ten years later, I got my license and the car became mine. It was the late 1970s of Jimmy Carter and the oil crisis, with lines for gasoline stretching for blocks. It was a time when locking gas-caps were sold to prevent one's hard-earned tank of expensive gas from being siphoned away in the night. I was seventeen, a teenager in working-class Long Island; the Falcon was fifteen, already looking rather quaint. We cruised around in the company of other early '60s' cars, GTOs and Mustangs and Camaros, the Falcon being the lightweight, little-engined sister to these dual-carbed, neutral-dropping, drag-racing muscle cars.

Our Falcon was well cared for, but cars age quickly, cosmetically more than mechanically. My mother, ever sensible, has no great affection for objects and would have happily traded it in, but my father, who had arrived in America from Hungary in 1957, with terrible memories of the Holocaust and more recent recollections of the violent bleakness of communist Europe, insisted on keeping it. Ostensibly, he was going to get his driver's license and the car would then be his. And, he invested objects with deeper meaning. This was the first car he had bought here in America—a part of the American Dream.

So the car wasn't traded in; it came with me to college, and then on to graduate school. It spent a year with me in the East Village, back when this now-gentrified neighborhood was a graffiti-covered district of crack houses and storefront galleries. My then-boyfriend, smitten by the retro coolness of listening to baseball games on an authentic AM radio, volunteered to make the daily search for alternate side of the street parking.

Together, the Falcon and I moved up to Boston. By then there were far fewer cars with shiny chrome and big round taillights. Everyone wanted to go for a ride in the Falcon, even though it had no air-conditioning in the summer, iffy heat in the winter, and the sort of doubtful brakes and steering that kept it in the right-hand slow traffic lane. The blue paint was faded, the fenders were rusty, but the car had style. No matter how dully mundane I felt, in the Falcon I was the Driver of that Cool Car. The Falcon was the car we drove away in when I got married. We had a newer car, faster, more reliable, but the Falcon was the festive, special occasion car.

A couple of years later our first baby was born. We bought books on baby proofing, put safety plugs in the outlets, lowered the water temperature below scalding. We looked at the Falcon and saw a metal steering wheel, hard dashboard, dubious brakes, and no shoulder belts. We found a new owner for the Falcon, someone who'd take care of it and keep it running, a nice guy with his own acetylene torch.

We gave the Falcon away to him in the end. Not because we could not sell it, but because the money would not be what it seemed worth to us. In its condition, it would sell, optimistically, for about $1500 or less. This seemed like an indignity to so personal a vehicle. Without money, the transaction seemed an adoption rather than a sale.

As a hammer extends the strength of the arm, a car is often seen as an extension of the legs. Yet a car is also an extension of the skin. As we inhabit our skin we inhabit our cars, surrounded by a rubber, glass and metal automotive skin that is both protective and expressive. Blemished with dents, rust marks, and scratches, it is an interface between the outside world and the inner self.

As we drive, our car changes our sense of personal space. Much of the process of becoming a driver is learning to be one with the car, to shift your perception of your own perimeter to the space around your vehicle. To know how wide you are, to judge what spaces you can squeeze into, to maintain the right distance between you and the car in front.

The car is a great equalizer; it makes every driver powerful and dangerous, yet it also signals individual diversity. We read car choice the same way we read clothing choice. A car can signal taste, money, or their lack. The Volvo wagon signals eastern liberal intellectual; its dented beat-up diesel disguises a love of all things organic and generations of Ivy League wealth. The Porsche driver in wrap-around sunglasses allows himself to be admired as he revs his car at the stoplight; some see him enmeshed in an aura of wealth and animal magnetism, others see him as the embodiment of the midlife crisis.

Driving, too, is expressive. The car revving impatiently at the stop sign has its aggressive driver, quick to take offense and with a chronic need to be first. The car that flinches at oncoming traffic, even though everyone is well within their proper lane, reveals the nervous reflexes of its frayed and jumpy driver.

When a car works perfectly, doing exactly what it's supposed to do, we experience it as pure machine. But when it acts imperfectly, choosing to do some things and not others, it becomes an almost autonomous agent, a seemingly sentient creature with emotions, desires and intentions of its own.

I negotiated with the Falcon when it exhibited its own preferences for speed, for direction. It required work to anticipate its likely reaction to my wishes. If I wanted to go faster I could not just depress the accelerator, because it would likely react by stalling. Instead, I needed to slowly add pressure to the gas pedal, letting up when it felt resistant and listening for a shifting of gears, then slowly, carefully, pressing down some more. Do I smell gas? Is there a hole in the gas line? Is the whole thing about to explode?

The Falcon stalled in the rain, and as it grew older, it stalled on any damp, low-pressure day, teaching me to become, like the pilot of a small plane, attuned to the meteorology of frontal systems. As I walked to work in the pouring rain I could think about the irony inherent in having a car that did not like to go out in bad weather.

Moving from New York to Boston I felt lonely and missed the casual interchanges of New York street chat. Yet, in the Falcon, I was in a friendlier Boston. People waved and smiled, and at stop lights they'd lean out of their windows to say something like, "How old is that car?" or "We had a car like that when I was a kid," or "Hey, I like your car!" With the Falcon, the city became an urban village of friendly strangers.

In 1994 I began a Web site for the Falcon. The Web was new, and I was studying its possible futures. When the Web arrived, people started making home pages, publishing their vacation photos and lists of their favorite movies and the details of their children's births. I read these pages avidly: What image of themselves were people trying to present on these pages? Who did they imagine as their audience? What motivated them to write these private stories in a public domain? As for writing anything about myself, I was torn between wanting to participate in this new medium and feeling too reserved to write anything personal. Creating a home page for the Falcon allowed me to participate in the personal and informal side of the web, while remaining at least partially hidden behind the persona of my car.

The Falcon's home page included commentary about its life as a car and links to information that a car would find interesting—other Falcons, car repair shops, articles about Boston drivers. Anticipating today's Web journals, it featured a frequently updated diary. Some of the entries depicted a "day in the life" of the Falcon:

> *Aug. 4, 1994*
> Today I sat by the curb.

> *Aug. 12, 1994*
> Still sitting at the curb. Hope we go for a ride later.

The Web journal also provided links for readers to send messages to the Falcon.

> *Dec. 19, 1994*
> Thank you to everyone who wrote expressing concern for my health. I'm feeling much better now. I have a new ignition switch and my heater has been refurbished.

Sometimes the Falcon departed from the mundane to wax philosophical:

> *Oct. 23, 1994*
> Went to see *Pulp Fiction* last night. Found it disturbing—the image of all those eviscerated cars serving as VIP seating in a trendy LA club continues to haunt me. The ethical questions are so complex. I know I've been very lucky, and that most cars my age have long since been junked. But the long slow rust of the junkyard is natural—and something about the sight of those polished, empty bodies bothers me. I know I won't run forever. Would I accept the strange immortality of the theme cafe? I don't know.

The Falcon's homepage played with questions of online identity. The Falcon had a clear personality: it was old, a little cranky, and insecure about its appeal, vacillating between feeling like a proud antique and a rusting junker.

Today I drive a BMW, a little 3-series station wagon. I have car seats for kids in the back; there are juice boxes stashed in the trunk, along with leftover beach shovels and odd pieces of toys. I like it in spite of its Germanic austerity. Unlike Japanese cars, which, like tiny Japanese apartments, have innumerable clever little places to stash things, the BMW has been carefully designed to store almost nothing. There is nowhere to put CDs or your handbag. One little pocket in the door has room for a couple of maps, and there is a little shelf in the dashboard to store one pair of sunglasses, and that is it. You are supposed to be focused on the road, on your role as the driver, and on your synergy with this fine machine.

Yet the BMW is not mute. It speaks of being a soccer mom's car, a car that says these kids in the back seat are to go to college, and that the grocery bags in the trunk are filled with colorful organic produce and imported olive oils, and that no coupons were clipped to buy them. This is an unremarkable message in our neighborhood of Saabs and Volvos, of ski racks and private beach parking stickers. I am no longer married, so the front passenger seat is usually unoccupied, and I stow everything there. It runs and handles beautifully. It is, as they say, "the ultimate driving machine." It does what I say. I feel no need to negotiate with it; I do not feel that it has a will of its own. No one stops to comment. New, it is a commodity with nothing to distinguish it from the many other BMWs in this city. Its biography is yet to be written.

I don't know where life has taken the Falcon. Until several years ago, I used to see it around town. Maybe it has once again changed hands. Or maybe the guy who adopted it went back home to New Hampshire.

Judith Donath is Director of the Sociable Media Group at the MIT Media Lab.

1964 Ford Falcon   **161**

The subject of passage ritual is, in the liminal period, structurally, if not physically, "invisible." . . . The transitional-being or "liminal *persona*" is defined by a name and by a set of symbols. The same name is very frequently employed to designate those who are being initiated into very different states of life. . . . We are not dealing with structural contradictions when we discuss liminality, but with the essentially unstructured (which is at once destructured and prestructured) and often the people themselves see this in terms of bringing neophytes into close connection with deity or with superhuman power, and what is, in fact, often regarded as the unbounded, the infinite, the limitless. Since neophytes are not only structurally "invisible" (though physically visible) and ritually polluting, they are very commonly secluded, partially or completely, from the realm of culturally defined and ordered states and statuses. Often the indigenous term for the liminal period is, as among Ndembu, the locative form of a noun meaning "seclusion site" (*kun-kunka, kung'ula*). The neophytes are sometimes said to "be in another place." They have physical but not social "reality," hence they have to be hidden, since it is a paradox, a scandal, to see what ought not to be there!

—Victor Turner, *The Forest of Symbols: Aspects of Ndembu Ritual*

THE SYNTHESIZER

Trevor Pinch

It was the sound that first drew me in. What was a police siren doing in a university common room during the annual "Freshers Fair"? The various university clubs had set up their stalls and we, the "freshers" (the British name for freshmen or first-year students), were prowling around looking to join the clubs of our choice. It was 1970—the tail-end of student radicalism. I made a beeline for the "Soc Soc" (the Socialist Society) stall. But I followed my ears to the source of the siren.

On a nearby table, surrounded by a small group of what we would today call "geeks," was a space-age machine. It was a contoured wooden box covered in arrays of colored knobs, controlled by a matrix panel of little pins. Sticking out from the box was a small joystick not unlike a radio-control for a model airplane. The sounds came from the box and seemed to be correlated with the movements of the joystick, which a young woman was manipulating with great skill. "Welcome to the Imperial College Electronic Music Society" proudly proclaimed a banner. I later learned that the woman was called Lindsay and that the box was a VCS-3 electronic music synthesizer. It was one of the first commercially produced synthesizers, made in London by the EMS company. It was a cheap version of its much more famous American cousin, the Moog synthesizer. Unlike the Moog it had no keyboard and no wires—different sounds were set up by connecting the different modules such as oscillators, amplifiers, and filters via the matrix pins.

The geeks were happy for me to play with the box, twiddle some knobs, and experiment by putting the matrix pins in different places. I asked, "What do you do in the electronic music society?" "We meet once a week and sit around and listen to electronic music," I was told. That sounded boring, almost as bad as the Marxist Study Group. But what the heck? I signed up.

The sitting around part did turn out to be terminally boring. The music, with its repetitive electronic timbres, was often mesmerizing, but the atmosphere was all wrong—geeks, loudspeakers, and a fancy tape recorder all in reverent silence. The worst was that there was no sign of Lindsay and her synthesizer. The next year, coincidence changed that. Lindsay moved into my communal house with its druggy music scene. We played guitars, bongos, alarm clocks, whatever, and tripped out. We commandeered Lindsay's synthesizer and played our guitars through it. We made jungle sounds, war sounds, a Vietnam psychedelic rhapsody.

Lindsay was unhappy. She had a very personal relationship with her synthesizer, which even had a name, Vickers. She spent nearly all her time playing with Vickers, making weird sounds into the early hours of morning. It was personal exploration for her—person and machine somehow evolving together into a new identity. When we borrowed her synth she felt Vickers was being abused. We were spoiling that delicate relationship she strove for, in tune with the idiosyncrasies of her instrument—never for her a machine—and its own peculiar sounds. One night we kidnapped Vickers: we needed the synth for a raucous session of music making. The next day Lindsay was close to tears when she told me she was moving out. No more loaning of Vickers; she just couldn't do it. It wasn't fair to Vickers.

I had to have my own synth, but even the cheapest synth was too expensive. Our communal budget was fifty pence a week each for food, and the VCS-3 cost over 300 pounds. What to do? The answer was obvious—build one. As a boy I had built radios. I soon graduated from tinkering with the mandatory crystal set to one- and two-tube shortwave radios. My pride and joy had been a shortwave radio called an R-1155, which I had picked up in an army surplus store. It had been stripped from a World War II Lancaster Bomber and had the words "Eager Beaver" etched across its giant dial. In my darkened bedroom I listened to Morse code and imagined I was bombing Berlin. With its huge tuning knob, a tiny movement of the hand switched you from Voice of America, to Radio Berne, to Radio Moscow. In between the stations was ambient radio sound—Morse bleeps morphing into howls of static amidst the weird booming of radio beacons trying to jam the stations you weren't supposed to listen to. If you sent letters to stations informing them when, where, and on what frequency you had listened, you were rewarded with postcards known as QSL cards. Much to my dad's alarm, I once received one in the mail from Radio Havana! "That's communist propaganda, son," he warned me.

Part of my radio hobby was the magazines—*Wireless World, Radio World, Electronics World,* and the like. I browsed them endlessly for new projects. The 1973 January edition of *Wireless World* had exactly what I was looking for—the Wireless World Synthesizer. I was soon ordering components by mail order, searching the cheap electronics stores on Edgware Road and Tottenham Court Road for elusive components. The metal work was all done at night in the physics laboratories of Imperial College, where I was still a student. My synthesizer had forty-five knobs and a joystick. Rather than matrix pins it used wires and plugs to set up sounds in a poor man's version of the Moog patching system. It even had a rudimentary sequencer that allowed me to program six different sounds in succession which would repeat endlessly. The transistors were always burning out—but repairing it was almost as much fun as playing it. I soon recognized the smells of different burning transistors and could almost feel which component was on its way out. I used an old organ keyboard that I had wired up with resistors to control the sound. The organ keyboard was impossible to tune so I abandoned tuning altogether and found that it was more interesting playing in my own invented scales or nonscales. I had a two-track tape recorder—the plan was to make my own electronic music by laboriously recording one sound at a time.

As time passed our communal living dissolved. The synthesizer moved with me to my garret apartment. Some nights I couldn't get anything interesting out of the synthesizer and then there were those magical nights when it seemed every new sound was a source of inspiration. I pictured the sounds coming out of the ether, like the radio stations I had listened to. A tiny movement of a wire or knob could make a huge difference. Filters were imperfect and the stray capacitance from my hand changed things. The reverb unit—built around a real spring—made interesting sounds if I shook its case. Broken modules were sometimes sources of frustration, but with experimentation I found they could produce even more interesting sounds. Wires straddled my synthesizer as each new botch and fix started to take on a life of its own. Closer and closer I got to the essence of electronic sound—no longer interested in making tapes, I just wanted to experience new sounds, to find the elusive combination of timbres that would enable transcendence. I escaped into my own world of sound. Was it music? I no longer cared. At last I started to experience what Lindsay must have felt—I was living with a machine and it was becoming part of me.

But as the patches got more complicated and the smell of solder too strong, it was harder to find the sounds I wanted. I started to hear differently; yesterday's sounds would no longer suffice, and even silences became rich with electronic timbres. I had to stop. It was too hard, too weird, and too lonely.

I withdrew slowly from the synth and came back from my isolation. I found new friends; a new career as an academic loomed. My time alone with the synth had served as a rite of passage. Now, the synthesizer became a mere curiosity for other people to explore. Once in the 1980s, one of my sociology students, who composed electronic music, discovered my old synth. We played long into the night, both amazed that in the digital age, it still had so much to offer. But as my soldering skills faded and I came to know my synth only from the outside—not the inside—I became reluctant to fire it up. When I moved to the United States and a new job at Cornell University in the 1990s, my synth moved with me. Now I lived near where Bob Moog (a Cornell grad) had invented the first synthesizers. I soon had a new passion—writing about the history and sociology of the synthesizer. My own synth sits in my basement to this day: silent and unnamed, but not forgotten.

Trevor Pinch is Professor of Science and Technology Studies and Professor of Sociology at Cornell University.

To get to the idea of playing it is helpful to think of the *preoccupation* that characterizes the playing of a young child. The content does not matter. What matters is the near-withdrawal state, akin to the concentration of older children and adults. The playing child inhabits an area that cannot be easily left, nor can it easily admit intrusions. This area of playing is not inner psychic reality. It is outside the individual but it is not the external world. Into this play area that child gathers objects or phenomena from external reality and uses these in the service of some sample derived from inner or personal reality.... [Thus] in playing, the child manipulates external phenomena in the service of the dream and invests chosen external phenomena with dream meaning and feeling. [And] there is a direct development from transitional phenomena to playing, and from playing to shared playing, and from this to cultural experiences.

—D. W. Winnicott, *Playing and Reality*

MURRAY: THE STUFFED BUNNY

Tracy Gleason

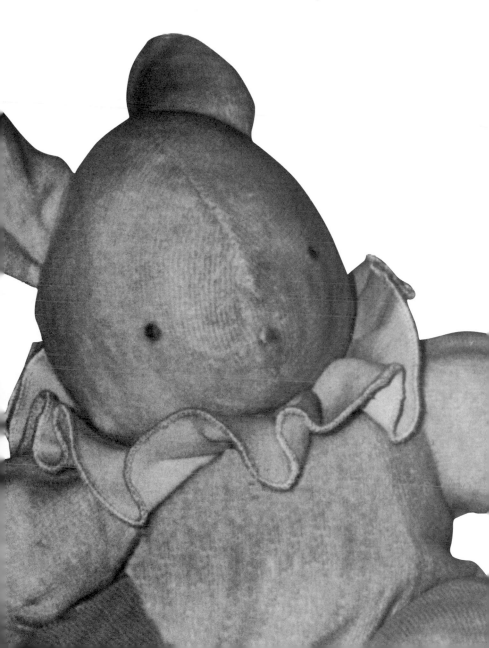

I had just begun my graduate research on imaginary companions, including children's animated stuffed toys, when my little sister, Shayna, was born. Like many little girls, Shayna received a host of stuffed bunnies in her first two years and quickly became mistress of a bunny menagerie. A student of scientific categorization, my father named each bunny according to its distinguishing characteristics. The smallest was Mini Bunny, and the two bunnies with clothes were named Big Jacket Bunny and Little Jacket Bunny. A Mama Bunny came with Baby Bunnies #1 to 4. A bunny with a soft cotton collar less than half-an-inch wide was named Collar Bunny.

Collar Bunny was for many months simply one of the menagerie. He was long on comfort and short on personality, just a stuffed bunny about 10 inches high, with floppy arms and legs, a big white head, and smallish ears that stuck straight up. His fur, sewn in stripes the shades of lightly decorated Easter eggs, did not distinguish him from the rest of the pastel objects of Shayna's world. He had a small plastic rattle inside his body, and when he sat, the stuffing in his arms made them stick out to the sides. He came to be valued for this ability to rest stably in a sitting position, a characteristic that made him a welcome guest at tea parties.

When Shayna was two, she saw a children's video for which her aunt and uncle had written the lyrics and music. A little girl in the video carried a large stuffed bunny named Murray with her everywhere. Shayna routinely acted out portions of the story, singing the songs and faithfully reproducing the blocking of each of the scenes. In the course of these activities, she chose Collar Bunny to play the part of Murray. Eventually, her interest in the video waned, but the idea that a little girl could love a bunny never did.

My friends and colleagues, who know of my interest in imaginary, personified objects, believe that I did everything in my power to encourage Shayna's relationship with Murray. Although I did not do so intentionally, I confess that Murray has been an endless source of fascination for me. He is my research personified in soft, velvety fabric. For me, Murray is better than a developmental psychology textbook. I see through him into Shayna's imagination. His ability to comfort, entertain, and amaze my sister delights me as a manifestation of our tendency to embody character inside fluff and fabric. When Shayna is upset, I watch as Murray dries her tears, and I am somewhat taken aback to discover that I, too, am comforted by his presence.

As a budding preschooler, Shayna grew in personality and power. And so did Murray. He became as important at playtime as he was at bedtime. Shayna would throw him up in the air and lift him in long leaps down the hallway, proud that Murray could push the envelope of bunny behavior to include ceiling-high jumps. In Murray I could see Shayna's pride in all of *her* new skills, like dressing herself and hopping on one foot and telling a silly joke. Soon Shayna could control and distinguish the words and actions of her real self, the role she took in play with Murray, Murray's "real" self, and the role she gave Murray in play with her.

When Shayna went off to nursery school, preschool policy dictated that Murray could not follow. Consequently, Shayna pretended that Murray was attending a different preschool, and she identified the church in which it was housed. Stressed by their separation, Shayna gave Murray a host of special powers. He developed Boing-De-Boing Eyes that allowed him to see through barriers of all kinds, around corners and across miles. He could see what Shayna was doing no matter where she was, and he always knew when she was coming for him. His Boing-De-Boing ears allowed him to hear Shayna speaking no matter how far apart they were. He could fly, magically transporting himself through space and time to be by her side—in spirit, if not in body.

When Shayna began kindergarten Murray developed new competencies. He and Shayna began communicating in the Bunny language, thus elevating their discussions beyond the comprehension of our parents. In order to keep them informed, Shayna gave Bunny language lessons, complete with tape recordings of vocabulary and worksheets for grammar. At school Shayna was a student; at home she was the teacher. As she learned to read, write, and spell English words, Shayna taught her mother these same skills in Bunny language.

As Shayna mastered literacy, the lessons in Bunny waned. The Bunny language still exists, but now it is the official language of Bunnyland where Murray attends elementary school. Bunnyland is a utopia of peace and prosperity, with festivals every Sunday and on alternate Wednesdays. It has provinces and capitals and a complicated topography, and we depend on Shayna to keep us apprised of Bunnyland's current events.

References to Murray's life both here and in Bunnyland provide Shayna with an entrée into adult conversation, and Shayna is happy to tell his tales. Murray's exploits give Shayna transitional, albeit imaginary, topics for the dinner table. Shayna uses Murray's experiences to forge a new kind of relationship with adults, one that is no longer solely based on her need for their nurturance, but rather is founded on common interests. When I visit I always ask about Murray's latest adventures, because I know they will reflect Shayna's current hopes, interests, and fears.

Murray shows signs of love and age; his jumpsuit is starting to tear a little over his bum, and his rattle is visible through worn patches and no longer makes any sound. A turn in the washing machine with something red made him decidedly more pink than white. At times when he is tossed aside—in favor of Barbie, say—at these moments, as he lies on the floor with his arms and legs akimbo, the simplicity of his being becomes apparent. Yet, when I find him on the floor, I feel compelled to pick him up and sit him in a more comfortable position, perhaps placing a book nearby in case he gets bored. I know his brain is polyester fill and his feelings are not his but my own, and yet his Boing-De-Boing eyes see through me and call me on my hypocrisy. I could no more walk past Murray as he lies in an uncomfortable position than I could ignore my sister's pleas to play with her or the cat's meows for food. Here, Murray has nothing to do with intellect and everything to do with love.

The adults in Shayna's life see Murray according to their own desires. My father thinks of Murray as a Puck-ish sage: a bunny with a great sense of humor, who is wise beyond his years. To him, Murray is the kind of guy who would have fun watching football but also has an opinion on Kant. Shayna's mother sees Murray as love personified, a being who would accept and forgive in the face of any transgression. For my husband, Murray is a kindred spirit. Each year he finds in Murray an oasis of calm in the chaos of Christmas morning (they are both Jewish).

For me, Murray has many faces. As a sister, I am grateful to him for bringing me closer to Shayna. I speak for him (always a high, squeaky voice appropriate for a small bunny) and can make my sister laugh. I hug him and treat him as a person, and my sister beams with pleasure. As a researcher, he has given me a ringside seat at the performance of Shayna's imagination, even as I remind myself that in fact it was she, as his creator, who bought me the ticket to that seat.

Finally, Murray teaches me about myself. When I think about him I can sense how willingly I blur the edges of fantasy and reality, and how we all choose imagination as a forum for practicing our social skills or safely experiencing powerful emotions. I animate Murray into giving me what I need, even as I know that he is an assembly of cotton and stitches.

I have two photos on my desk. One is of my husband, and the other is of Murray. In his photo, Murray's body almost fills the frame. He is seated next to Mini Bunny, and he has his arm around her. His head is cocked to one side, and he is holding a daisy in his lap. I know that my sister arranged the picture, but in my heart I feel it captures his personality to a T.

Tracy Gleason is Associate Professor of Psychology at Wellesley College.

It is the world of words that creates the world of things. . . . Man speaks, then, but it is because the symbol has made him man. . . .

Symbols in fact envelop the life of man in a network so total that they join together, before he comes into the world, those who are going to engender him "by flesh and blood"; so total that they bring to his birth, along with the gifts of the stars, if not with the gifts of the fairies, the shape of his destiny; so total that they give the words that will make him faithful or renegade, the law of the acts that will follow him right to the very place where he *is* not yet and even beyond his death . . .

In order to free the subject's speech, we introduce him into the language of his desire, that is to say, into the *primary language* in which, beyond what he tells us of himself, he is already talking to us unknown to himself, and, in the first place, in the symbols of a symptom.

—Jacques Lacan, *Écrits*

THE *WORLD BOOK*

David Mann

How they decided, I do not know. Maybe in a moment of grace they sensed my need and chose to help me. Maybe they were just "keeping up with the Joneses," as one did in those days. Did the neighbors have a Philco, a Eureka, and a box of books, necessitating ours? I do not know how it happened, but as improbable as it looks from the distance of these years, my parents bought for us the 1952 edition of the *World Book Encyclopedia,* and, in doing so, literally gave me a world.

I came from a family of very few words. For us, living was a private matter, best tended to in silence. Speech was less a gift than a liability. In the culture of our clan, true conversation, opening oneself to another's point of view, could never have taken hold. To proclaim, "Here's what I think; what about you?" would have amounted to an act of civil war, a threat to the sovereignty of individual experience and an invitation to the other's scorn. "Better to be silent and thought a fool, than to open your mouth and prove it," warned one of the few maxims I recall hearing as a child.

It sounds oppressive, but I do not believe my family intended that we live this way. As best I can tell they had not shunned the larger culture, nor intentionally banished it from our camp. They had just arisen outside its reach, in a time and place and circumstance where little beyond the King James Bible and the iron skillet had yet found a fertile niche. The dust bowl of Oklahoma had spawned my parents, starved their spirits, and forced them out into the world where they clung to each other, to memories of simpler times, and to their silence.

My family huddled—yet we traveled, too. We moved household nearly every year, tethered to my father's military service. Travel can sophisticate a family, give them a chance to learn the meaning of their ways, open the world to them and show them their place within it, teach them humility and grace. Our travels, though, seemed to isolate us more, to thicken the scar of our alienation. I remember as a child, my first day in England, watching a television ad for dog food "rich in doggy vitamins." We stood in the parlor of a modest B&B, proper but not prim, as a few guests sipped their afternoon tea and my parents inquired about a room. Given the prosaic matter of his sales pitch, the television announcer's tone, so intelligent and poised, struck me as funny. His way of saying "vitamin" (sounding like "cinnamon") rang in my Yankee ear as silly. I laughed aloud. The room fell silent. A spotlight of shame pinned me to the creaking floor. I can still feel my father's rage, my mother's mortification, the guests' indifferent huff, the TV spaniel's relief as she sniffed the bowl. I no longer recall my punishment—most likely a slap to the face and exile to a distant room. It was always safer to be alone.

My earliest memories bear a feeling of separateness from the human world—and not only among strangers, but within my family as well. Ironically, what saved me from despair was that they let their silenced offspring stray. In my early years we lived mostly in rural places, where even a small child could drift untended for long stretches of the day. I climbed in trees, dawdled in streams, poked at bugs, played with clouds and gravity and angular momentum, and found comfort in this world outside human commerce.

My family caricatured the naivete that had bred them. It was the 1950s, in the USA. Our culture was young, childish, really. The world beyond our shores was frightening and dangerous, a *terra incognita* prone to dictatorship and war. Other cultures we feared as primitive, hostile, or both. Mau-Maus and Maoists. Other governments tortured and lied. Ours was a kindly Father Who Knows Best. Over There, ideas could madden crowds and kill. Here, our own ideas seemed to protect us with their insular magic. A wide-eyed faith in progress drew us forward despite our mistrust of change. Though legally free, we cowered in conventionality, mumbling prayers to science in the callow faith that it would save our world. In a stolid but uneasy balance of centrifugal hope and centripetal fear we reeled. This culture leached into me as a child. I felt it cringing in my bones, commingling there with other urges that it opposed but could not neutralize—a playfulness with the familiar and a curiosity about all that lay beyond. What I needed was a guide, but none appeared. Indeed, my people rather disapproved. "Don't mess with that!" "Be still!" "Shut up!" "Don't ask stupid questions," they admonished.

I do not remember when the box came. I must have been a toddler still. I do remember how the books looked in it, because I packed and unpacked them so many times over the years that followed. They stood at attention, each crimson spine wore a swath of blue, lined in the same gold paint in which the letter identifying each volume had been stamped. Out of their box the volumes resisted opening (like their owners), but once ajar they released a scent of glossy paper and halftone ink that to this day recalls feelings of amazement, challenge, and comfort. Gentle masters, the books offered their secrets freely and never shamed me for inquiring. Had they filed out of their box and marched onto the ceiling I would have tried to follow them. I carried them with me throughout whole days of wandering. They became my interpreters, my models, and my guides.

At first the books showed me only pictures, of other places, other times, other treaties with the elements: a water clock, an Archimedean screw, a doll sewn from a sock, another strung from empty spools, a sandaled Phoenician inscribing the precursor of our letter "H" in clay, a Ubangi girl, a spiral galaxy, an igloo, a boy on a hillside gleefully racing his homemade kite into the wind. These were pictures that I could both identify with and wonder at. They were windows of possibility opening onto a world wider than the one I knew but where I felt I could belong. Like an index to my mind, these images still appear to me when I search for words, much as they taught me words as I pored over them as a child.

The *World Book* was my Rosetta Stone. Its pictures came to life in my mind, parsed into nouns and danced through grammar to the music of verbs. By the time I was four it had taught me to read. Not through my family but through these volumes language became a part of me, the book of the world opened to me and I myself opened to the world as I might otherwise never have done.

As a physician and psychoanalyst, I have had many teachers, but the *World Book* was my first, the one that taught me how to learn. Today I help others through their own, similar transitions, from alienation to belonging in the world, from chaos to conversance. Often, like the *World Book,* my comradeship is silent. And, like the *World Book,* I try to be available with images and words for the experiences that have silenced those who seek my help. I must work to grasp their feelings; I often fail (the book of my soul resists opening), but I am grateful for the chance to try.

David Mann teaches at Harvard Medical School and practices privately in Cambridge, Massachusetts.

If in our earliest development we have been able to transfer our interest and love from our mother to other people and other sources of gratification, then, and only then, are we able in later life to derive enjoyment from other sources. This enables us to compensate for a failure or a disappointment in connection with one person by establishing a friendly relationship to others, and to accept substitutes for things we have been unable to obtain or to keep. If frustrated greed, resentment, and hatred within us do not disturb the relation to the outer world, there are innumerable ways of taking in beauty, goodness, and love from without. By doing this we continuously add to our happy memories and gradually build up a store of values by which we gain a security that cannot easily be shaken, and contentment which prevents bitterness of feeling. . . . Then we are actually capable of accepting love and goodness from others and of giving love to others; and again receiving more in return. In other words, the essential capacity for "give and take" has been developed in us in a way that ensures our own contentment, and contributes to the pleasure, comfort, or happiness of other people.

In conclusion, a good relation to ourselves is a condition of love, tolerance, and wisdom towards others. This good relation to ourselves has, as I have endeavored to show, developed in part from a friendly, loving, and understanding attitude towards other people, namely those who meant much to us in the past, and our relationship to whom has become part of our minds and personalities. If we have become able, deep in our unconscious minds, to clear our feelings to some extent towards our parents of grievances, and have forgiven them for the frustrations we had to bear, then we can be at peace with ourselves and are able to love others in the true sense of the word.

—Melanie Klein, "Love, Guilt, and Reparation"

THE SILVER PIN

Susan Rubin Suleiman

For a long time, I thought of it as a precious thing: a flower pin, long and slender, the sculpted leaves spreading on both sides of the stylized petals, with two symmetrically placed pearls in the middle. My mother wore it on the collar of her black dress in the photos we posed for before we left Hungary. It was in the spring of 1949, a few months before we crossed the border into Czechoslovakia. I still recall the session with the fancy photographer, who came to our house and had me leaning against doorposts in casual, girlish poses (I was nine years old). He also took more formal pictures of my parents and me, including the one of my mother in her black dress, sitting at a table with her arms resting on an open book. Her left hand, very white and smooth, stands out against the black of her sleeves. Her head is slightly tilted, and around her mouth there plays a slight, sweet smile. She looks kind and beautiful, her eyes shining, her dark hair a halo—an elegant, still young woman of leisure. One would hardly believe, looking at her manicured hands, that a few years earlier she had been working as a maid in Budapest, hiding from the Nazis with false papers. My father and I had been there, too.

On the back of the photo, which I now hold in my hand, its edges slightly frayed, is an inscription in flowing black ink: *Sok szeretettel, Lilly*—with much love, Lilly. She had sent this photo to her mother, my grandmother *Rézi nagymama,* who had left Hungary the previous year with my uncle Lester, her eldest son. They were allowed to take the train, no need to walk across the border—that was before the Communist regime in Hungary cracked down on emigration. Rézi was in New York City, where we eventually joined her. Her youngest son, my American uncle Nick, who was doing well in the shoe business, had set her up in a one-bedroom apartment on York Avenue, not far from the mayor's mansion, in a tall brick box that was the latest thing in those years, with doorman and elevators, and air conditioners visible in all the windows as you looked up at the blank facade. It was in her apartment that I was introduced to the wonders of television—we watched Molly Goldberg and Milton Berle religiously, and *Dragnet,* too ("Just the facts, Ma'am"). How strange America was, and how green I felt!

Curiously, I have no memory of my mother ever wearing that pin after we came to the United States. She held on to it, that's certain—I have it in front of me right now, on the desk next to the photo, and keep glancing at it as I write: the pearls are slightly yellowed, and if I look closely I see many imperfections on their surface. The sculpted leaves, too, show signs of decrepitude, dotted with small gaping holes that were once filled in with glittery stones—"not diamonds," the jeweler told me recently when I took it to him for his opinion. The holes look almost as if they were there on purpose, as if the designer had wanted to alternate empty spaces with filled-in ones. It's an old pin, graceful in shape and commercially worthless. "Enjoy it, it's pretty," the jeweler said.

So why didn't she wear it? Was this modest relic of postwar Budapest unworthy in her eyes? (I think my father bought it for her shortly before the photography session, as a sign of prosperity and survival). Or was it perhaps associated with a country, and a city, that she had no desire to remember? She had lost most of her extended family in 1944, deported with the help of the Hungarian government—she never spoke about those uncles and aunts and cousins, and I have no memories of them since almost all of them lived in the provinces, far from Budapest; but when I was writing my book *Budapest Diary,* I made a pilgrimage to the city where she was born, where some of the family had lived. She spent her summer vacations there as a child. I can't even begin to imagine what it felt like for her to learn, at war's end, that all those people were dead.

Some immigrants retain their ties to the old country. I know Hungarian Jews in Boston who still refer to Budapest as "home," decades after they just barely escaped being shot into the Danube by Hungarian Nazis— that was quite the sport in the fall of 1944. Some left the country soon after the war, like us; others waited until 1956, fleeing when the borders became temporarily crossable after the failed revolution. They all started going back for visits in the 1960s and 1970s, when "goulash Communism" made life in Hungary quite pleasant again, especially for Hungarians with American passports and dollars. My uncle Lester returned to Budapest every summer for more than twenty years, right up to his death. Communism or no, the Gerbeaud pastry shop on Vörösmarty Square still served the best sour cherry strudels, and you could dine outdoors on chicken paprikash with *nockerli* at the Duna Corso restaurant on the bank of the Danube, late into the night. Not to mention music and theater, the best in the world, according to him.

My mother had no truck with such nostalgia. She never went back to Budapest and reminisced about her youth in that great European capital only if I pushed her hard, with photos spread out before us. "The Gellért baths, I went there often when I was courting," or: "Do you remember our Sunday hikes in Buda, when you were little? You loved the cog railway." Generally, she sought advancement and novelty, not memories. She had a talent for small talk with strangers, and within a few weeks after we arrived in New York she had established several outposts of acquaintances in the neighborhood. I especially recall the children's clothing shop on 86th Street near York Avenue, where she would go to chat with the owners in a mixture of German, Hungarian, and broken English as she looked for outfits for my baby sister. She didn't hesitate to ask them for a discount, given our status as new immigrants. Often I felt embarrassed when I went with her, especially when she pushed me forward to translate for her or when she started telling people she had just met about her most intimate concerns: her worries about money, her anxiety about our future, her doubts about my hair! ("You must do something about your hair," the refrain of my adolescence!). It was around that time that I began to feel she and I had nothing in common.

It occurs to me that maybe she did wear the pin in America, and it is I who have blocked it from memory. Was I ashamed of her for not being American? Was the pin, which I had thought splendid and precious in Budapest, now merely a reminder of foreignness?

After the first two years of struggle in New York, we moved to Chicago, where I went to high school—another displacement, another round of feeling like an outsider. For a short while, I had an intense friendship with a girl I thought of as the perfect American. She lived in a large frame house on the North Side, with her parents and a sister and brother—he was older, already in college, but came home for the holidays. At Christmas, they put a big tree in the middle of their living room and went caroling in the snow. I don't remember what her mother looked like, but I recall wishing my mother were more like her—she never yelled, never nagged her daughter about her appearance. She was calm, not excitable, and embarrassingly familiar with strangers. After a few weeks, my friend and I drifted apart, or maybe she snubbed me. Today, I recall only the yearning I felt to be like her, to have a family like hers.

Looking back on this now, I realize how desperate I was to be an "insider," not different, just like other

Americans. And how ashamed I must have been of my immigrant mother, who never learned to speak English properly and never learned to speak calmly. But the drive for assimilation came from her as well: in a curious way, I was fulfilling her desire by wanting to have little to do with her. Success in school was my escape, my chance to leave her and foreignness behind. I was offered a scholarship to Barnard College in New York, and jumped at it. My mother was happy, too, knowing I was in a fancy school. Sometimes a new acquaintance would notice an accent and ask me about it, but most people I met in college thought of me as a girl from Chicago. I had a little black dress for parties, and my hair was finally in shape.

Back in Chicago, my father gradually made his way to a job he was proud of, as the executive director of a Hebrew day school. In the summer before my senior year, ten years after we had left Budapest, he died of a heart attack at the age of forty-nine. It took me a long time to mourn for him, but that is another story. We gave up our apartment in Chicago and sold its meager furnishings. My mother and my little sister, who was nine-years-old, lived for a year in New York, then moved to Miami Beach to be near my uncles and aunt. Meanwhile, I spent a year in Paris after college (generously financed by Uncle Nick), then moved to Cambridge, Massachusetts, to start graduate school at Harvard. Another displacement, another promotion.

I visited my mother once or twice a year: she treated me like a dignitary, parading me to family and friends. Rarely did a visit end without some outburst on my part. I had no patience with her; it was clear that we would never understand each other. I had adored her as a child in Budapest, but that time was very far away. A few years later, after I got married and became a mother, too, there were no more outbursts; the gap between us, however, persisted.

It makes me feel sad and ashamed—with a different shame, not the shame I felt as a teenager—to realize how little I valued her. But there is anger there, too. If I was incapable of feeling love for her—or of expressing love, which in a sense is the same thing—was it not her fault as well as mine? I tell myself that she was tactless, that she spoke too loudly, that she was interested only in the superficial signs of success.

Yet, others did love her. After almost twenty years of widowhood she married again, and her new husband doted on her. He was a retired dentist, Hungarian-Romanian, a widower—they got along well together, a real couple. When she became ill a few years later, he took care of her; when she died in 1988, aged almost eighty, he mourned her as if they had been together a lifetime. After her death, we kept hearing from people who had known her—she had been the belle of Lincoln Road, one old lady told us. She was fun to be with; she had a thousand friends.

My sister and I often talk about her now: she was impossible, yes, but she was brave and energetic, too, and she had gone through a lot.

We inherited her photos and her few pieces of jewelry. I got the old photos from Hungary, many with inscriptions on the back. Among them was the picture of her wearing the silver pin, so elegant and beautiful. The pin also came to me, along with a delicate gold orchid pin she had acquired in America. I put that one in my jewelry box; the silver pin disappeared into a jumble of old trinkets in a drawer: an antique belt buckle given to me by a French friend many years ago, broken or unmatched earrings, watches that no longer ran. Devalued, like my mother in America? Yes, but not thrown out—lying dormant.

The gold orchid, when I wear it, often reminds me of my mother; but it is simply a pretty object, carrying no strong emotion. The silver pin evokes bruises and ambivalence, emotional knots difficult to untangle. When

I dug it out of the drawer, it was nearly black with grime. I tried dipping it in jewelry cleaner, but it still remained dull and dark, so I took to it with silver polish and managed to get it to shine. It's quite pretty, as the jeweler said. I pinned it on a black jacket I wore a few weeks ago. I haven't worn it since then, and don't know when I will again. But it has moved to the jewelry box on top of my dresser—I suppose that's progress, of a sort.

Susan Rubin Suleiman is C. Douglas Dillon Professor of the Civilization of France and Professor of Comparative Literature at Harvard University.

Objects of Mourning and Memory

Superman, by definition the character whom nothing can impede, finds himself in the worrisome narrative situation of being a hero without an adversary and therefore without the possibility of any development. A further difficulty arises because his public, for precise psychological reasons, cannot keep together the various moments of a narrative process over the space of several days. Each story concludes within the limits of a few pages; or rather, every weekly edition is composed of two or three complete stories in which a particular narrative episode is presented, developed, and resolved. Aesthetically and commercially deprived of the possibility of narrative development, Superman gives serious problems to his script writers. . . .

There is nothing left to do except to put Superman to the test of several obstacles which are intriguing because they are unforeseen but which are, however, surmountable by the hero. . . . But this resolves nothing. In fact, the obstacle once conquered (and within the space allotted by commercial requirements), Superman has still *accomplished something.* Consequently, the character has made a gesture which is inscribed in his past and weighs on his future. He has taken a step towards death, he has gotten older, if only by an hour; his storehouse of personal experiences has irreversibly enlarged. *To act,* then, for Superman, as for any other character (or for each of us), means to "consume" himself.

—Umberto Eco, "The Myth of Superman"

DEATH-DEFYING SUPERHEROES

Henry Jenkins

I bought the comics on the way to the hospice. They were selected hastily and even then, I felt guilty about the time it took. I was looking for something banal, familiar, and comforting at a time when my world was turning upside down. I read them intermittently as my family sat on deathwatch, my experience of the stories becoming interwoven with our common memories and the process of letting go of my mother. Retreating from the emotional drama that surrounded me, I found myself staring into the panic-stricken eyes of a young Bruce Wayne, kneeling over the newly murdered bodies of his parents. I had visited that moment many times before, but this time, our common plight touched me deeply.

I am hardly the first to draw such connections. In his essay, "The Myth of Superman," Umberto Eco describes the monstrous quality of the superhero who is not "consumed" by time, never grows older, but always cycles through the same kinds of experiences, never moving any closer to death. Eco approaches this question formally, describing how the iterative structure of comics creates its own kind of temporality, which he contrasts with the always already completed action of myth or the unique events of the novelistic: "He possesses the characteristics of timeless myth, but is accepted only because his activities take place in our human and everyday world of time."[1] The fan boy in me wants to point out all of the exceptions and qualifications to Eco's claim, starting with the fact that a whole generation of revisionist writers has sought to reintroduce death and aging into the superhero universe. The images of aging Batman and Superman duking it out in Frank Miller's *Dark Knight Return* come immediately to mind, but most of these books came after Eco's essay was published and they might well have been responding to his argument. Regardless, Eco misunderstands that for serious comics readers, the same events may unfold again and again, but there is something distinctive about each issue. Mastering those distinctions is part of what separates fans from more casual readers. From time to time, the franchises build up such complex histories that they need catastrophic events—such as the Crisis of Infinite Earths—to wipe the slate clean again and allow a fresh start. Yet, such reservations aside, Eco's formal analysis hit on a core psychological truth.

One could understand the reading of comics as entering into a psychological space that both denies death and encourages a nostalgic return to origins. Most of our stereotypes about comics fans start from the idea of arrested development—that is, the idea that the fans have somehow sought to pull themselves out of life processes and to enjoy the same kind of timeless existence as the guys and gals in tights. I want to suggest the opposite, that in their own way, both as texts and as artifacts, comics become reflective objects that can help us think about our own irreversible flow toward death. In short, this is an essay about what it means to consume and be consumed by superheroes.

At night, I am frequently so tired that I fall asleep too fast if I try to read prose. I find that I can maintain consciousness, however blurry my perceptions, long enough to make it through an issue of a comic. I find something energizing in the shift between text and images and in the larger-than-life stories so many comics tell. Comics are the site of enormous diversity, innovation, and experimentation, but nowhere else in popular culture can you find the same degree of continuity. Superman, Batman, Wonder Woman, and Captain America have been in, more or less, continuous publication since the 1930s or early 1940s—always fighting for truth, justice, and the American way, despite generations of readers and writers growing up, growing old, and yes, dying in their company. There have been enormous variations in how these characters have been interpreted across generations. There have been dramatic shifts in styles, successive waves of revisionism, various stabs at relevance or topicality. Yet, you can go away for decades on end, find your way back to a DC comic, and get reintroduced to the protagonists more or less where you left them. It is often this hope—of rekindling something we once felt—which draws us aging comic fans back to some of these titles. It is almost as if we would lose something important within ourselves if watching Batman stalk across a darkened alley or Spiderman swing from building to building no longer made our hearts beat a little faster. Indeed, when I think of my own history with comics, so much of what I remember are iterative events, the routine patterns of my heroes rather than specific storylines.

My earliest memories of comics bring me back to my mother. When, as a fourth or fifth grader, I would stay in bed with a fever, my mother would go to the local druggist in search of Coca-Cola syrup, which according to Southern folklore, was supposed to have remarkable curative powers. In that era, before specialty shops offered reliable subscriptions, she would return carrying an armload of comics, selected from a large spinning rack at the center of the store. She brought more or less what she could find, so sometimes she would return with a selection of kids' comics (*Baby Huey, Donald Duck,* or perhaps *Archie*), other times with *Classic Illustrated,* and still other times with DC superheroes. I came to associate comics with the sound of my mother's voice singing me to sleep or her hands feeling my forehead. And I suspect that's why I return to them now at moments of stress.

It is hard to remember when superheroes first entered my life. I suspect it must have been 1966, the year that Batman first appeared on television. I was seven or eight. The series rapidly became an obsession among the neighborhood kids. One of my aunts had given me a recording of the theme music, which my playmates and I played at full volume, bouncing up and down on the bed, biffing and powing each other, and tumbling backward into the pillows. My mother had given me an old leotard, sewed a cape and cowl, and cut me a batarang out of plywood. We didn't always understand what we saw. Once we heard an announcement for a forthcoming appearance of King Tut and thought the announcer had said King Duck, and so we spent a week battling it out in the backyard with web-footed foes, before discovering that there was this Ancient Egyptian guy. Who knew? My father would peer out from behind his newspaper, expressing mock horror to have discovered that Batman and Robin had died that very day—frozen to death in a giant snow cone or in some other death trap. And every time, I would fall for his joke, bursting into tears, since I could never make up my mind where Batman stood in relation to the dividing line between fantasy and reality.

Some of the boys in the neighborhood formed a superhero club. I remember swearing an oath of loyalty over a stack of comics in my tree house. We each chose the persona of one of the members of DC's Justice League. The guy who lived across the street was unnaturally big and strong for his age and was quickly cast as Superman. The kid next door was small, wiry, and fast on his feet and he became the Flash. I had tired of Batman by this point and aligned myself with the Green Lantern for reasons long forgotten. We each spent our weekly allowance on comics and would pass them around. I have reread some of the stories of the era, only to be disappointed. We fleshed out the superhero personalities through our play and most of what I recall so fondly wasn't to be found on the printed page.

Most of those kids have disappeared from my life and rarely enter my thoughts. Recently, one of them tracked me down on the Internet. When we got together, we talked in big breathless gulps about boyhood days and then suddenly, silence fell upon us. We looked at each other blankly as if we were suddenly confronting not the boys we were but the middle-age men we had become, and we ended the evening early. Neither of us has called the other since.

As I sit down to write this, I am haunted by a curious memory—one of my few memories that centers on a unique event rather than a pattern of repeated experiences. It is early summer and I am sprawled out on the floor of my family's cabin in the North Georgia Mountains coloring a picture of Batman as my mother watches television across the room. I have spent most of the day thrashing about in the water pretending to be Aquaman and am now awaiting bed, when a news report interrupts the show my mother is watching to tell us that Robert F. Kennedy has been shot. Why are my memories of my mother's tears over Bobby Kennedy's death so firmly linked to my memories of superhero coloring books? Is it because at such a moment—which would have come when I was in sixth grade—I suddenly understood the line that separated the plots of the campy television series from the harsh realities of adult life? What had it meant to me to see my mother crying and not know how to comfort her?

You could say that what drew me to comics the week my mother died was nostalgia—which can be described as a desperate hunger to return to a time and place that never really existed, a utopian fantasy through which our current longings get mapped onto the past. Comics were comfort food, like the Southern cooked vegetables my mother used to fix for me when I came home for holidays. Yet, these comics offered me little comfort. I hurt every place my mother had ever touched me and found myself unable to separate out the comics from the memories they evoked. If comics brought me back to boyhood, then they brought me closer to a period in my life when my mother's love had been the most powerful force in my life.

I was surprised that I didn't stop reading comics while my mother was dying, but as an adult I had been reading comics every week for years. I returned to comics sometime in my mid-thirties—searching for something I couldn't name at the time. A few years later, when I was diagnosed with gout, I found myself drawn even more passionately back into the world of the Flash and the Green Lantern. A growing recognition of my own mortality drew me into the death-defying world of the superheroes, who, unlike me, never grew older and never had bodies that ached. For me, the comics work as a reverse portrait of Dorian Gray: They remain the same while my body ages and decays. As such, they help me to reflect on the differences between who I am now and who I was when I first read them.

I am telling too simple a story because as an adult, comics kept coming in and out of my life. There were various attempts to get my own son engaged with comics, all doomed to failure; or the way the release of the Batman films rekindled my passion for that character; or my periodic raids on comic shops to examine some title that a student brought to my attention. Although there are huge gaps in my knowledge of any given character, and whole comic series that came and went without my knowledge, I never really left comics. However, it took me a while to admit that I wasn't just wandering into comics shops now and again to see what was new, but that I was going there every week and coming away with bags full. As an adult, it took me a while to come out as a comics fan.

Even though my own work on fan culture had debunked many stereotypes about science fiction fans, there was a side of me that still believed clichés about middle-age comic book readers. If I have an origin story for my passion for superheroes, I also have an origin story for my fear of becoming a comics fan. It begins in Tom's smelly basement when I was in seventh grade and had decided I was too old for comics and ready to move on to more mature reading matter, like *Mad* or *Famous Monsters of Filmland*. Tom was a somewhat pudgy kid who lived down the street from my grandmother and we became friends initially out of geographic accident and emotional necessity; his house was a place to go when I wanted to escape being cooped up with someone who was constantly complaining about her aging and ailing body. Tom had just moved to Atlanta from Michigan. At a time when all of my other friends were committed to DC, Tom read almost exclusively Marvel. We would sit in his basement and rummage through a huge mound of yellowing comics, reading late into the night by flashlight. By then, reading comics was something you did to escape from the controlling gaze of moms. Tom's two cocker spaniels snorted somewhere in the dark void around us. Tom's basement smelt of dog breath, [fossilized] poop, and mildew—things that make you faintly uncomfortable when you are a boy but which grow in your memory with each passing year. As an adult, I am uncomfortable with the degree to which boys who held onto their comics into adulthood bore a stigma of arrested development. It was as if we had never left the basement or the tree house, still hanging out with the boys, still imagining what it would be like to be an adult, and that basement-like atmosphere pervades most of the dark, subterranean, and clubhouse-like shops where comics are most often sold.

Tom was perhaps the first comics collector I knew. I would meet many others, each waging their own campaign against death and decay, protecting their treasures with plastic bags, acid-proof cardboard backing, and steel boxes. Consider the case of two undergraduate friends who were both comics collectors, both guilt-ridden Catholics and both named Mark. One of the

Marks was a square-jawed fellow who wanted to be Clark Kent. He wasn't just dull, he was desperately dull. For him, memorizing as many facts as he could from Superhero concordances was one of the ways he could bring his corner of the galaxy more fully under his control. Years later, when I began to seriously collect comics, my wife bought me some reference books at a used book sale. When we examined them closely, we discovered Mark's name scrawled on the inside front cover. I'm not sure what surprised me the most, that Mark had finally gotten rid of those books or that my interest in comics had grown to the point where I saw value in owning these books in the first place. The second Mark took me to his apartment and showed me an entire room full of steel boxes, containing thousands of individually bagged comics and creeped me out with a speech about how his comics would be safe and secure long after he was dead. Years later, I visited him in Brooklyn and sure enough, he still had all of those boxes of comics and many more. By that time, however, I wanted nothing more than to sit up all night asking him for recommendations. The mausoleum had become a library.

For all of that, collecting comics wasn't terribly different from collecting any other kind of book. But there is a key difference. Unlike, say, leather-bound books, comics were never made to last. They were printed on cheap paper with bad ink, and the assumption was that they would be read and discarded. But no one ever thought that people would still be reading them decades later any more than one imagined holding on to old newspapers. Superheroes may be invincible, but comics rot. What makes old comics valuable for collectors is that so many of them have been destroyed. Every mom who threw away her son's comics increased the fortunes of those who were lucky enough to hold on to theirs. Many fans spend their entire life and much of their income trying to recover the issues they had once discarded so casually. And so, fans become preoccupied with the challenges of preserving their collections, with forestalling their ultimate destruction.

To her credit, my mother never threw away my comics. She took them up to the lake house and left them in a drawer. Over the years, they were literally read to death. Young visitors would paw through them with peanut butter-covered fingers. The staples came undone and pages would come off when you tried to read them one last time. The humidity made the pages more and more waterlogged and mildewed. The sun bleached the lurid covers as they were left for too long lying on the window ledge. And in the end, not a single one of the superhero books made it past my adolescent years. The *Classic Illustrated*s were more expensive than the rest—and came with the aura of high culture—so mom treated them as sacred and eternal, not unlike the way she dealt with *National Geographic* magazines. They are the only comics from my childhood that I still possess. I still recall how many Jack Kirby books got ripped up when a Boy Scout troop was rained-in for one weekend at our cabin, but I still loan out my comics to my students rather than worry about keeping them in pristine condition. I have refused to take that last step into fan boy culture. For the moment, I am more interested in reading and sharing comics than in keeping them out of harm's way. I know nothing lasts forever; you are better off really enjoying the things you love while you can.

These are some of the thoughts that passed through my head as I sat on my deathwatch. I put aside Batman, not ready to face young Bruce's angst, and turned instead to Spiderman, only to find a comic storyline that dealt with the memories stirred up by the anniversary of Uncle Ben's death. Eco is right that superheroes don't move closer to death. In fact, they move further away from it with their origins often bound with trauma and loss. Yet, death defines the cycles of their lives. Almost all of the comics I brought to the hospice dealt—at least in part—with childhood trauma and loss. If comics provide youthful fantasies of empowerment and autonomy, they do so by severing the ties between the superheroes and their parents. Batman takes shape in Bruce Wayne's mind as he vows vengeance over his parents' tombstones. Superman's parents send him away from a dying planet. Peter Parker, not yet aware that with great power comes great responsibility, is too self-centered to stop a crook, allowing him to escape and kill his Uncle Ben. What separates the villains from the heroes isn't the experience of loss, but what they did after that loss, how it shaped their sense of themselves and their place in the world. Some were strengthened by loss, others deformed.

Most of the literature of childhood has emotional violence at its heart. Through fiction, we expose children to the real life forces we seek to shelter them from, almost inevitably the death of or separation from one's parents. In comics, these events do not occur one time, but crop up again and again; these images of death and mourning define characters' identities. In the months that followed my mother's death, I found myself returning, almost involuntarily, to memories of her final days, the way that a tongue seeks out and presses against a loose tooth just to see if it still hurts. I came away with a new understanding of why the superheroes hold onto their grief; they can draw upon it as a source of strength. At one point in my life, I read those stories to learn what it was like to have the power and autonomy of adulthood. Now, I read them to see how to confront death and still come out the other side. They helped me realize the common human experience of loss and recovery.

The comics of our childhood are impossible to recover. Even if you hold on to your comics, the stories on the page are not the same ones you remember, you find something new and different each time you come back. In my case, the death-defying superheroes helped me find a way to hold on to my mother while letting go.

Henry Jenkins is Director of the Comparative Media Studies Program at MIT.

Through photographs, each family constructs a portrait-chronicle of itself—a portable kit of images that bears witness to its connectedness. It hardly matters what activities are photographed so long as photographs get taken and are cherished. Photography becomes a rite of family life just when, in the industrializing countries of Europe and America, the very institution of the family starts undergoing radical surgery. . . . Photography came along to memorialize, to restate symbolically, the imperiled continuity and vanishing extendedness of family life. Those ghostly traces, photographs, supply the token presence of the dispersed relatives. A family's photograph album is generally about the extended family—and, often, is all that remains of it. . . .

The force of a photograph is that it keeps open to scrutiny instants which the normal flow of time immediately replaces.

—Susan Sontag, *On Photography*

THE SX-70 INSTANT CAMERA

Stefan Helmreich

The Polaroid SX-70 camera, introduced during the 1970s, was a folding chrome-and-leather single-lens reflex camera that looked like a cross between a tiny, trapezoidal accordion and a collapsible robot toy. It delivered instant color photos, framed in white plastic borders, in just under 1.5 seconds. Once outside the camera, in the light, the pictures took about a minute to develop fully, ripening from an initial turquoise haze into a creamy colorful lucidity, a process one could watch through the transparent Mylar membrane covering the swirl of chemicals that would constitute the photograph. In the time it took for SX-70 pictures to materialize, experimentally inclined people like myself would sometimes smudge and smear the colors beneath the Mylar—an activity more famously engaged in by the artist Lucas Samaras, who took many Polaroid self-portraits and then mutated his likeness into fantastic shapes.

My grandfather, Howard G. Rogers, a chemist with only a year of college, at Harvard, during the Depression, invented some of the pliable molecules inside Polaroid's instant color film. His key creation was a molecule called a dye-developer, a compound that fused image dyes to photographic developers, allowing instant color film, in effect, to embed its own darkroom chemicals. His dye-developer molecules sat in limbo at the bottom of the photo frame of each unexposed Polaroid photo card and, with the snap of the SX-70 shutter, would be squeezed up into the picture plane by rollers inside the maw of the camera. As pictures emerged from the SX-70's tight mechanical jaws, they made a wonderfully distinctive noise, something like: *Zt-ZzzzT.* For some, the one-minute wait that followed was too much; when the film exited the camera, these impatient folk would wave the photo in the air to hurry along its development (This gesture—which my grandfather informed me was completely useless—was commemorated in the 2003 hit song "Hey Ya" by the rap duo Outkast, in which one line enjoins people on a dance floor to "Shake it like a Polaroid picture"). Growing up, I was always curious about how SX-70 film worked, and from time to time, my grandfather would narrate me into the microscopic, millisecond world within the layers of a Polaroid picture.

The problem before my grandfather had been this: to get three color dyes—cyan, magenta, and yellow—to express the complementary colors to which they corresponded: red, green, and blue. A primary requirement was that different dyes not bleed into each other. Another was that variable rates of dye formation be controlled. Within the time that an instant color photo came into being, events had to unfold in a tightly compressed time sequence. The problem required understanding events on extremely small spatial and temporal scales.

My grandfather's idea was to fuse dyes and developers into one megamolecule. Effectively joining these ingredients would allow the elements of photography to be squashed into a compact space—and, more, would enable the instantaneity of instant photography itself. This scheme, however, went against a prevailing wisdom that believed it risky to put dyes and developers into close proximity. But Edward Land, my grandfather's boss, was committed to the notion that when confronted with an obstacle, one should consider doing the opposite of the expected.[1] My grandfather took this wisdom to heart. In his Patent #2,983,606, granted on May 9, 1961, dye developers are described as key components of "novel processes for forming monochromatic as well as multicolor pictures by transfer and reversal practices wherein a single reagent is utilized for the formation of a negative image as well as a positive image of said negative."[2]

Reflecting on his invention later in his life, my grandfather said, "When an idea like this comes, that you're sure is good, it spreads throughout your body. I felt intoxicated, but more 'all there' than usual—almost as if I were a giant."[3] This language triggers memories of my grandfather chatting with me over the dinner table, shrinking me down to the size of an atom, so that I could rub shoulders with molecules and then zoom back out to look at a family photo taken with the SX-70.

All of our family pictures were taken with Polaroid film. In-laws sometimes grumbled that the colors were not as vivid as they might be, which always sent my grandfather into a distracted accounting, storing up complaint and commentary for his next visit to the lab. Ours was a kin group wed not just to family photos, but also to a family photo *technology*. It was incumbent upon us to be loyal to my grandfather's attempts to get his colors right, which meant that we also had to be dedicated to Polaroid products. In a way, the SX-70—a cryptic abbreviation of "special experiment seventy," a code name Polaroid used for the realization of absolute one-step photography—made of our family an experimental laboratory. And while my grandfather clearly enjoyed his time with his five children and five grandchildren, particularly at the lakeside cabin he and my grandmother purchased in Maine with Polacolor profits, he often seemed preoccupied. Years later, he reflected in print on preoccupation, distraction, and inspiration:

> I became more and more impressed with the power of the subconscious. . . . If you put good input into your subconscious, that is, carefully observed results and carefully thought-out analyses, and let some good hard facts into your subconsciousness, along with the need to know the answers to some

problems or the need to invent the way out of some difficulties, then sometimes further focusing and work wasn't as helpful as just a little time, or a change of scene, or a stimulus of another sort [which] would sometimes bring the answer.[4]

The family, was, I think, for him, "a stimulus of another sort," a technology for jostling his subconscious. Elements of daily life at home became a playful experiment—from his fascination with engineering tiny poached eggs with fractionated yolks to his proclivity for taking stereoscopic pictures of me and my cousins at moments when we were embarked on some particularly three-dimensional enterprise, such as learning to sail.

In other words, my grandfather's work became part of the family's play. My mother—growing into an adult in the psychedelic sixties—modulated my grandfather's fascination with color into her own stirrings of chemicals in the paints she used in her watercolor paintings. In the mid-1970s, I made birthday gifts for my grandfather that made fun of the sciences of imaging. One present, a favorite, described an imaginary invention that I dubbed "the image inverter." It turned images upside down so that one could see them the way the eye actually receives them. Another consisted of a manual for a camera with no lens. Always ready for a laugh, and to consider the unexpected, my grandfather found these takes on his professional work hilarious and displayed them prominently.

My cousins and I began smearing Polaroid pictures at around the same time as Lucas Samaras. My grandfather gave us advice on getting the best results and was always eager to watch his invention unfastened from its original aim. As we transformed family photos, our extended family was itself in transformation. In the 1960s and 1970s, our parents' generation had swerved away from the middle-class Catholic-Protestant model of my grandparents. I was born hours before my parents were married. One of my mother's sisters sidestepped marriage and Christianity altogether, moved into the Maine woods with a back-to-the-land mountain man, and joined him in raising their kids in the Jewish tradition. My grandfather greeted all these transformations with equanimity. My grandmother grew into a Catholicism that became ever more, well, catholic. The SX-70 pictures from this period reveal traditions morphing and mutating.

Later in life, after retirement, my grandfather would glide into occasional reveries about new inventions he wished to realize. Sometimes, the oxygen he took for his emphysema would intoxicate him, and he would describe such things as edible dyes that, once ingested, could accentuate color perception. In what has become a piece of family folklore, Polaroid scientists were once summoned to his bedside during one of these rhapsodic episodes, to determine whether this now-renowned chemist might be hatching new, counterintuitive, but perhaps effective ideas for color photography. According to these corporate visitors, this was not the case. But rather than seeing this story as one in which my grandfather takes a detour away from himself, I view it as revealing a reversed but true image of my grandfather, much like the image that bounces off the interior mirror of an SX-70 at the last moment before the exposure of a photograph. I see my grandfather's reveries as an attempt to reverse engineer—with the aid of the oxygen tank that he, after all, controlled—the feeling of intoxication he associated with invention; maybe his occasional flights of fancy were a direct sounding of the subconscious he found so intriguing. I like to think that he was taking us on a tour of the kinds of worlds sited within SX-70 film, a domain in which the rules of reality were understood at a higher degree of resolution, where molecules caught up in the representation of familiar people, places, and things revealed themselves at the most microscopic level to be mirrors of our ever-changing selves, developing and transforming

Stefan Helmreich is Associate Professor in the Anthropology Department at MIT.

May 9, 1961   H. G. ROGERS   2,983,606
PROCESSES AND PRODUCTS FOR FORMING
PHOTOGRAPHIC IMAGES IN COLOR

Filed July 14, 1958   3 Sheets—Sheet 3

Image—Receiving Layer,
Support

Rupturable Container Holding Processing Composition

11
15   17
54
52
80
64
62
80
74
72
40

Blue—Sensitive Emulsion
Layer Containing Yellow Dye Developer
Spacer Layer
Green—Sensitive Emulsion
Layer Containing Magenta Dye Developer
Spacer Layer
Red—Sensitive Emulsion
Layer Containing Cyan Dye Developer
Support

F I G. 9

INVENTOR.
Howard G. Rogers

BY
Brown and Mikulla
and
Stanley H. Mervis
ATTORNEYS

The old saying: "We bring our lares with us" has many variations. . . . The house is not experienced from day to day only, on the thread of a narrative, or in the telling of our own story. Through dreams, the various dwelling-places in our lives co-penetrate and retain the treasures of former days.

And after we are in [a] new house, when memories of other places we have lived in come back to us, we travel to the land of Motionless Childhood, motionless the way all Immemorial things are. We live fixations, fixations of happiness. We comfort ourselves by reliving memories of protection. Something closed must retain our memories, while leaving them their original value as images. Memories of the outside world will never have the same tonality as those of home and, by recalling these memories, we add to our store of dreams; we are never real historians, but always near poets, and our emotion is perhaps nothing but an expression of a poetry that was lost.

—Gaston Bachelard, *The Poetics of Space: The Classic Look at How We Experience Intimate Places*

SALVAGED PHOTOGRAPHS

Glorianna Davenport

I stare at the first photograph that I have pulled out of a small cardboard box labeled "Glorianna to make copies." It is a picture of my father in his youth by a lake with a dog. I never knew my father had a dog. Three years ago, I promised my siblings that I would digitize a large collection of memorabilia—images and videos. For this, I recently added a scanner to my image-processing setup at our cranberry farm. My promise still unfulfilled, guilt is balanced with the anticipation of new discoveries. As I continue to muse, the image of my father is transformed into bits.

The small cardboard box is filled with an unruly group of most-wanted photographs selected by my siblings as we painstakingly divided up family heirlooms after selling my mother's house in 1999. My mother did not choose to be part of these final decisions. She already had her fill of dealing with the remains of her family heritage ten years earlier, when the house my grandparents had built in the 1930s—in which we had spent our childhood summers, and into which my parents moved after my grandparents passed away—was consumed by a devastating fire.

The reaction of survivors to the sudden and near total loss of personally meaningful possessions can create a sort of frenzy. In our family, my mother and all of her children contributed to clearing and then to combing the site, carefully freeing larger and smaller artifacts from the debris of our lost home. For the two months following the fire, we each, at different times, searched the charred remains for anything that might be recoverable and threw the rest into a large dumpster.

When I arrived at the recovery scene, the smell of wet char was overwhelming. My older sister and her husband had turned a small space in the old library into a conservation lab. This room was the least damaged by the fire; however, the water that poured through the floorboards as the firemen fought the flames above had flooded the bookshelves and drawers, the boxes with letters dating from the early 1800s forward, and almost a century of family photographs. The image of my father that I have just scanned is one of the few surviving fragments of this twentieth-century collection.

The salvaging of paper artifacts requires a different kind of patience from that required by larger objects such as pianos, sofas, and china. With paper, the injury caused by heat, water, and falling debris are soon augmented by mildew if the paper is not properly dried. In this fire, the damaged pages numbered in the thousands; conservation required each page to be separated and dried before sorting into collection categories. That any paper trail of our family history exists today reflects the painstaking work of my older sister and brother-in-law.

The fascinations of our youth give shape to our future passions. As I begin my efforts to digitize my family's history, I realize anew that this box holds not only a collection of photographs but keys to many of my later life decisions. Photographs led me to cameras, and over the years the camera became an object I could think with. I could think about light and shadow, about composing the frame, and about what it meant to live in a certain way, to make decisions at many levels, and to document the world. I could think about the chemistry and the mechanics of editing. I remember the brown leather case of the Rolleiflex sitting on my mother's bureau. I can smell the chemicals that pervaded the photo shop on Madison Avenue. I don't know how old I was when I first looked into the lens and noticed the sight and sound of the cuh-chunk-click made by the shutter mechanism. The mystery of this mechanism continued to fascinate me for many years.

Slowly I separate a few large square black-and-white prints that have stuck together. Who is this strikingly beautiful woman wearing a luscious lipstick and bending over two young girls and a boy dressed in party clothes and wearing paper party hats? The youngest girl—possibly three—plays with beads. The older of the two girls, perhaps five, holds the hand of an older boy in a checked jacket and bow tie. Another boy stands in the distance. Only the older girl looks toward the camera: where are the others looking? What was the event? I think my mother may have taken photographs for other families; perhaps this photograph was not even taken in our home.

Other pictures seem more familiar. A laughing child is about to spray water across a familiar terrace; this is clearly me. A woman bounces a laughing baby on her knee; I easily identify my aunt, who later died of cancer in our home. An image of my father on a sailboat, a picture of my youngest sister acting as Mary in a school play—each image awakens memories and poses new questions.

Reaching into the box once more, I pull out a narrow album of Polaroid portraits, its covers long since lost, its pages formatted with cascading transparent envelopes held together by a plastic binding. This album used to sit on the piano in my grandparents' living room. Most of the portraits were taken by my grandfather. In a moment, I am back at the site of the fire: the buckled floorboards, the drenched piano, the charred Polaroid camera. And then just as quickly, the album becomes a prism through which I see that house and its objects before anyone imagined there would be a fire: the framed documents dating back to the founding fathers, the music chest, the way my grandfather hid modern electronic contraptions in two-hundred-year-old chests.

My mind returns to my grandfather and his Polaroid. By the time I was eight or nine, my grandfather had become mentor to my technological bent. For years it had been my grandfather's practice to take portraits— group shots of the family celebrating one of many summer birthdays, as well as individual portraits of each grandchild year by year, usually taken out on the grass "circle" in front of the house. My brother John sits next to a large pumpkin in the circle; my younger sister Sharon hugs a pumpkin and stares wistfully into space; my sister Ann sits on a large rock in Maine; my sister Susie is posed next to the sewing box she had won in a sewing competition. What do these images tell of the uneven, often difficult roads that lay ahead for each of us?

I do not remember exactly when, but one summer I discovered that the way to avoid being in the picture was to take the picture. My grandfather, an electrical engineer trained at MIT, loved gadgets and regularly purchased and experimented with the latest Polaroid cameras. I must have enticed him to let me borrow one. He seemed to enjoy teaching me about the nature of the lens and how best to frame a shot. However, he soon tired of my using up the costly Polaroid film packs for experiments that did not yield images that he found meaningful, and gave me my own 35 mm camera—a PONY IV—for my birthday. With this, I was able to control depth of field with focal length and shutter speed and could experiment more freely with composition and subject.

Several summers of intensive photographic activity ensued. Since sending photographs away for factory processing was expensive and did not allow maximum control over the image, my grandfather decided to set up a small darkroom in the bathroom across from his lab. This experimental operation irked my grandmother who—especially in the heat of the summer—could not tolerate the smell of the chemicals that overran the house. My grandfather prevailed: I spent the next two or three summers avidly recording the world in black and white, processing rolls of negatives, and exploring the disciplines by which one selects and then controls the printed image.

The editing function allows us to pick, choose, and modify content as we examine each image for its evocative value. These early explorations served me well in my adult career as I moved into film and interactive video, seeking a better understanding of how stories are made and shared.

The process of recovering from the house fire has brought me in touch with old lessons: not all documents are worth salvaging, but most are, and photographs are particularly valuable to later generations of a family, allowing them evidence to better reconstruct the tale of their past.

I will no doubt digitize all the images that were "most wanted," but when it comes to printing and framing the "story," I will be more selective. The editor in me, picking through the debris after the fire, is uniquely positioned. Having found the charred but no longer functioning Polaroid camera, I could choose, if reluctantly, to relegate it to the dumpster as I now can navigate through its images, developed, digitized, and jostling for position in some future album.

Glorianna Davenport is Principal Research Scientist at the MIT Media Lab.

I raised to my lips a spoonful of the tea in which I had soaked a morsel of the cake. No sooner had the warm liquid mixed with the crumbs touched my palate than a shudder ran through me and I stopped, intent upon the extraordinary thing that was happening to me. An exquisite pleasure had invaded my senses, something isolated, detached, with no suggestion of its origin. And at once the vicissitudes of life had become indifferent to me, its disasters innocuous, its brevity illusory—this new sensation having had on me the effect which love has of filling me with a precious essence; or rather this essence was not in me it *was* me. . . .

And suddenly the memory revealed itself. The taste was that of the little piece of madeleine which on Sunday mornings at Combray (because on those mornings I did not go out before mass), when I went to say good morning to her in her bedroom, my aunt Léonie used to give me, dipping it first in her own cup of tea or tisane. The sight of the little madeleine had recalled nothing to my mind before I tasted it; perhaps because I had so often seen such things in the meantime, without tasting them, on the trays in pastry-cooks' windows, that their image had dissociated itself from those Combray days to take its place among others more recent; perhaps because of those memories, so long abandoned and put out of mind, nothing now survived, everything was scattered; the shapes of things, including that of the little scallop-shell of pastry, so richly sensual under its severe, religious folds, were either obliterated or had been so long dormant as to have lost the power of expansion which would have allowed them to resume their place in my consciousness. But when from a long-distant past nothing subsists, after the people are dead, after the things are broken and scattered, taste and smell alone, more fragile but more enduring, more unsubstantial, more persistent, more faithful, remain poised a long time, like souls, remembering, waiting, hoping, amid the ruins of all the rest; and bear unflinchingly, in the tiny and almost impalpable drop of their essence, the vast structure of recollection.

—Marcel Proust, *Remembrance of Things Past*

THE ROLLING PIN

Susan Pollak

If I close my eyes, I can almost go back to my grandmother's kitchen. The fragrance of pot roast permeates the air, redolent with caramelized onions, potatoes, and carrots. I can see the golden lemon sponge cake, made with nearly a dozen eggs, just emerging from its worn silver Bundt pan. And I can smell the cups of steaming black tea with sugar. This was Grandma Tilly's healing elixir, which could soothe any pain and still the rivers of my childhood tears and adolescent rage.

A shaft of sun on the kitchen table illuminates the sugar bowl and the flowered, blue plastic tablecloth. The light reminds me of the serenity of Vermeer's interiors and of his women, completely absorbed in their domestic tasks. I see my grandmother in her apron, her hair the purest, softest white. She is legally blind but is holding her beloved rolling pin. Even though she can see only shadows, she is still cooking for us, baking the most delicious sweets.

To think about my grandmother, with her rolling pin and her fragrant kitchen, is to meditate on loss. She was the stable anchor in my life, mediating between an absent, depressed father and an irrational, erratic mother. Thankfully, she lived behind us so I could escape to her kitchen when I needed solace. I remember the sheer joy of climbing over the stone wall that separated our houses and bounding into her kitchen, feeling both free and deeply connected.

I was nine months pregnant with my first child when she died. Even though she was seriously ill, she was holding on, waiting for the birth of her first grandchild. We were hoping, praying, that she would be able to cradle the child in her arms. Every day felt like a race between birth and death. Tilly, a union organizer with a will of steel, seemed in control of her death.

The baby was breech, and my doctor informed me that unless it turned I would need a Caesarian section. Days before the baby's due date, my grandmother died. I spent the night weeping, mourning her loss and the fact that my children would never know her warmth and her kindness. During that night of grief, the baby turned, its head pointing down, ready to be born.

My grandmother has been dead for nearly fifteen years, but when I make cookie dough with my children I use her wooden rolling pin with its chipped red handles. I exert gentle pressure and roll the dough back and forth. I add flour and flip it over to the other side. This tactile ritual takes me back to the warmth of her kitchen, the aromas of her cooking, and the comfort of her presence. As I bake, I often tell my children stories about Grandma Tilly. The loss is still present but now bittersweet. I miss the comfort of her world, yet I am deeply grateful that she was such a presence in my life.

As I use her rolling pin and feel its texture and weight against my floured hands, I think of the hundreds of pies and cookies it helped create. It anchors me in the past, yet continues to create memories for the future. The object becomes timeless

Marcel Proust gives us deeper insight into the nature of the evocative object. In *Remembrance of Things Past* he describes an epiphany evoked by a madeleine, a small, scalloped cake: "dispirited after a dreary day with the prospect of a depressing morrow, I raised to my lips a spoonful of the tea in which I had soaked a morsel of the cake. . . . A shudder ran through me and I stopped, intent upon the extraordinary thing that was happening to me. An exquisite pleasure had invaded my senses."[1]

This humble cake set into motion Proust's masterpiece on memory and loss. His poetic understanding

of the power of the senses to evoke a state of consciousness is unmatched:

> But when from a long-distant past nothing sub-
> sists, after the people are dead, after the things are
> broken and scattered, taste and smell alone, more
> fragile but more enduring, more unsubstantial,
> more persistent, more faithful, remain poised a
> long time, like souls, remembering, waiting, hop-
> ing, amid the ruins of all the rest, and bear un-
> flinchingly, in the tiny and almost impalpable drop
> of their essence, the vast structure of recollection.[2]

Evocative objects can hold the "vast structure of
recollection." This is more than poetic construction—objects can have a profoundly healing function. The British
psychoanalyst D. W. Winnicott developed the idea of the
"transitional object." We think of the child's teddy bear
or the "blankie" as a link to the love and comfort of the
mother, but Winnicott also located the capacity for ten-
derness and caring in such objects. What is less known,
but germane to the purpose of this essay, is that they are
also the basis of symbolism and creativity: "In this way
I feel that transitional phenomena do not pass, at least
not in health. They may become a lost art, but this is
part of an illness in the patient, a depression, and something equivalent to the reaction to deprivation in in-
fancy. . . .[3] The object can hold an unexplored world,
containing within it memory, emotion, and untapped
creativity.

As a psychologist, I inhabit multiple worlds.
Through transference and countertransference, I have
a special relationship to the stories, dreams, and objects
of others. Working with my patients, I become both
translator and participant/observer of their inner landscape. When a case deeply engages me, the objects and
stories of others assume weight in my world, inhabiting
my thoughts and imagination.

The Case of Mr. B.

Mr. B., a fifty-year-old married man, entered treatment to work on an abusive and traumatic relationship with his parents. He was a novelist, but had been unable to write for a number of years. During the course of treatment, his father died after a long illness. The father, a distant, tyrannical alcoholic, never let his son know that he loved him. Although Mr. B. had written a number of books, his father had never made an effort to read them.

A number of months after his father's death, Mr. B. was visiting his mother. During the visit, they returned to the town where Mr. B. had grown up. Out of nostalgia, he looked for the bakery that made his favorite treat, a thin cake covered with chocolate and vanilla frosting, called a "half-moon." One of Mr. B's fondest childhood memories was of his father surprising the family with a box of these cookies. Remarkably, the bakery was still in business, and Mr. B. bought a box of half-moons for himself and his children. Because he had grown up in difficult times when money was tight, his own father bought day-old cookies, which were often broken and stale. Mr. B had never tasted the cookies either fresh or whole.

To his taste buds, there was something wrong about the moist, intact cakes. He saved them, waiting for them to become stale. After a few days, the texture was "right"—the frosting hard, the cake dry—and he could savor and re-create the lost tastes of his childhood.

Never underestimate the power of an evocative object. The incident with the cookie—the finding of a lost object and sharing it with his children—gave him access to the "vast structure of recollection." Entering this forgotten world of smell and taste was a pathway to new memories. Some symbolic essence of childhood had been recovered. As Mr. B. grieved, he was able to recover positive feelings about his father that had eluded him for the two years of our treatment. For months after the death he had experienced a profound terror. He had recurrent nightmares where he would look for his father in vain, searching in attics and basements, only to find the house in ruins, littered with shards of glass and shattered bricks.

The re-finding of the cookie corresponded to a turning point in Mr. B's grief. This sweet was a concrete and positive link to his past, an evocative object that was both sustaining and stabilizing. Mr. B. was able to recall acts of generosity and to develop a deeper understanding of his father's need to live in a drunken haze. He was able to tell his children stories about their grandfather. The cookie had become a gateway, connecting him to "enduring and faithful" memories. It became a way to integrate what was positive and "sweet" about his father. Slowly, with hesitation, Mr. B. began to write again; he began a novel about childhood.

Toward the end of *Remembrance of Things Past,* Proust makes a connection that Winnicott would wholeheartedly endorse: "Ideas come to us as the successors to griefs, and griefs, at the moment when they change into ideas, lose some part of their power to injure our heart."[4] Proust's stymied protagonist is able to give up his fruitless search for his lost mother, which frees him to act. As his pain is transformed into ideas and images, he begins to write.

My patient experienced a parallel process. When he had mourned, remembered, and worked through all that he had not received from his father, his rage and paralysis subsided and he was able to write again. Winnicott would agree that a return to health is also a return to creativity. The evocative object holds more than memory; it holds healing potential. We create our objects and are inspired by them. As I found with my rolling pin, and my patient with his cookie, the evocative object is transitional in the fullest sense of the word—it can bring together generations, anchor memory and feeling, and evoke attachments that have long been forgotten.

Susan Pollak is a Clinicial Instructor in Psychology at Harvard Medical School.

Whenever we are trying to recover a recollection, to call up some period of our history, we become conscious of an act *sui generis* by which we detach ourselves from the present in order to replace ourselves, first, in the past in general, then, in a certain region of the past—a work of adjustment, something like the focusing of a camera. But our recollection still remains virtual: we simply prepare ourselves to receive it by adopting the appropriate attitude. Little by little it comes into view like a condensing cloud; from the virtual state it passes into the actual; and as its outlines become more distinct and its surface takes on color, it tends to imitate perception. But it remains attached to the past by its deepest roots, and if, when once realized, it did not retain something of its original virtuality, if, being a present state, it were not also something which stands out distinct from the present, we should never know it for a memory. . . .

But the truth is that we shall never reach the past unless we frankly place ourselves within it. Essentially virtual, it cannot be known as something past unless we follow and adopt the movement by which it expands into a present image, thus emerging from obscurity into the light of day. In vain do we seek its trace in anything actual and already realized: we might as well look for darkness beneath the light.

—Henri Bergson, *Matter and Memory*

THE PAINTING IN THE ATTIC

Caroline A. Jones

The object in question is stock-in-trade for an art historian; the illustration shows it: an oil painting in the standard illusionist mode. Of course, as a painting (even a bad one), it is meant to transcend thing-ness altogether. We are meant to assemble the photons reflecting from this colored mud-on-canvas to see a grouping of people, apparently children of various ages, standing together in an indeterminate space. What becomes evocative about this object for a given viewer is unpredictable, but probably the expressions on the children's faces would provoke some thought: two girls smile, a boy conveys mock surprise, a small girl looks solemn, a baby screams. The dramatically different facial expressions on these children seem to seal each one into a separate world; none seems to react to any other. Instead, each face projects an image of an inner emotional state.

What is evocative for the "I" of this essay is much more specific than these general observations, because it was "I" who painted this picture more than thirty years ago. *Untitled* depicts my own brothers and sisters (and myself). A bit embarrassing, it hangs in my attic, relegated to a spot where only a few outside my family will ever see it.

Some might believe that I, as the artist, know all there is to know about this painting. But my claims are much more humble. Since I was the maker, I can limn rather precisely the boundaries of my own ignorance at that time. There will be certain things I can use from my professional toolkit as an art historian to interpret aspects of the painting that were not obvious to the girl who painted it. Likewise, there are aspects that become available to me *as* that person that would not be evident to any other art historian. What I want to emphasize here is the coruscating light that gleams between these surfaces—the practical problems firing up an art student's enthusiasms in 1972 and the strategies used by professional art historians to pry insights from objects in 2006. There is iridescence in this layered view. Interpreting the light relayed from one surface to the other produces the self-reflection characteristic of my present self.

Some art historians assert a quasi-positivist reliance on data—historical inventories, interviews, critics' writings, letters, sale records, as well as the visual layers of paint and canvas, preparatory studies, related works—to help argue for a single interpretation that will replace all other readings of the privileged object. This piling-on of evidence is more for the consumers of art history than for we who live within it; *we* are fired by inspired hunches, intuitive insights, strong viewings that help us make sense of "evidence" in the present. The evidence is necessary, and the positivist approach works for those who are uncomfortable about the leaps of faith necessary to tie facts into meanings. Indeed, on one occasion when I opened the floor to questions following a psychoanalytic analysis of the erotic space in Richard Diebenkorn's drawings, one audience member asked me scornfully, "Well he's alive, why don't you just *ask* him?" In the case of the present object, I possess all the omniscience my interlocutor might have dreamed of. Surely I can solve any puzzle about the *Untitled* painting at hand, because I have unparalleled access into the privileged intention of the maker. Surprisingly, what I have to report is that interpreting one's own work is no less complex than finding meaning in another's.

There are many concrete things I can tell you about this concrete thing, but they will not add up to an absolute truth. I can state the circumstances and struggles that attended the painting's production during my senior year of high school: how separate photographs and family snapshots were used as source materials (providing one explanation for the mix of expressions), how the background defeated me, how I ran out of patience and ideas and simply daubed a bosky blend of Thalo green and Cadmium yellow to suggest a vague, verdant setting. I can talk about the painting's trajectory as an object—how it survived years under a leaky roof in my parents' house, how creosote or some other roofing chemical dripped down the front of the canvas. How cleaning that off left a trace, a vertical line coursing past my brother's depicted ear. I can reminisce about where this unframed painting hung—over a particular couch in the family TV room, barely coordinating with the nubbly turquoise fabric—and how the people it depicted often sat on that same couch, changing more quickly than the painting would.

But the most extraordinary thing about this object is what I realized upon re-encountering it later as a professional art historian, something I had never considered as its maker: namely, that the image is *retrospective*. My age in this depiction (I am the girl at far left) is about eleven. When I painted it I was seventeen. This simple fact produces an anachronistic slippage between the denotative date of the picture (the ages of the children in the painting, their hairstyles, clothing, even postures allude to 1966) and the chronological date of the picture's completion (1972). Few viewers would find this remarkable, but it floored me when I recognized its obvious significance. The painting's convenient optic excludes a member of my family who was born in 1967—the retarded youngest sibling in this clan.

Thinking with this thing confirms one of my strongest convictions as a professional art historian— what the painter consciously puts into the picture is a tiny fragment of what can be gleaned by later viewers.

Intention is but a minuscule part of art's meaning; the aesthetic object has a larger, and often much longer, life. As it came back into *my* life, I was able to reflect on the omission of my youngest sister, who grew up with me, was and is adored, but still struggles to find a place in a family (and a world) predicated on competency and achievement. Even when I painted the picture, I already knew that the presumed trajectory for each of us—college, marriage, children, career—would not be likely options for this individual, this "special" sib.

Evocation is customarily sweet. We summon this word to describe emotions that tingle with nostalgia, savoring the subtle flavors of past pleasures and recalling childhood play. Here, evocation is more painful, as I am forced to acknowledge that I simply deleted my youngest sister to produce the image of an ideal phase of my family's life, before the traumatic event that would forever mark us as different, our familial business as never worthy of the happy-ever-after of all story lines. More interesting than the admittedly amateur painting itself is this strategic anachronism. Completely invisible to me then, the attributes of this repressive fantasy are obvious to me now. What interests me is the way such traces were nonetheless embedded in this painting by the adolescent who made it, coded messages in a bottle cast on the *bilderflut* for later viewers who might want to parse the patterns in the paint.

Far more than representational skill, the artist I aspired to be had to deploy the constituents of painting precisely *to evoke* a situation, without words, to tell a tale. Arguably, this has been the main job of pictures since the Renaissance. Photographic cameras, in this argument, merely codified a preexisting value system in which the perfectly composed image of a single moment prevailed over narrative change. Like the Christian icon that lies behind so much of Western aesthetics, the oil painting was intended to be an eternal object whose perfected, crystallized composition would never be disturbed by the vicissitudes of time. Time would even be phenomenologically banished from the pictorial realm—it would never be acknowledged as itself part of a viewing regime. Pictures in the Western tradition would be composed to convey a single event, the "pregnant moment" praised by Enlightenment aesthetician Johann Joachim Winckelmann as the highest form of art. But the painting under consideration, *Untitled* 1972, refuses the *prägnanz* of a single moment. That it does so indicates more than lack of skill. In other words, we might find a technical reason for the lack of internal unity to these depicted figures, for the obvious disparity in their emotional moments. Indeed, I've acknowledged one such technical reason—the multiplicity of photographs and snapshots from which the image was concocted. But attributing the lack of unity solely to this fact begs the question of why those disparate photographs were chosen in the first place. Such a solution fails to address the painting's *refusal* to be unified. The painting insists that the perfect moment is always already fractured, never unified in the first place. One preteen mugs, the baby cries, the little one refuses the obligatory camera smile. Only the oldest two—me and my older sister—appear to be *composed* for the camera/painting. Pictorial idealism fragments in the face of a reality it aims to signify.

The depictions of these particular individuals in some cases align with what I can tell you of their fates. The then-youngest girl is portrayed with the sense of sober moral judgment that would govern her adult practice as an acutely political feminist art writer; the oldest boy shows the goofy sense of humor that would win him the love of friends, family, and coworkers; the oldest girl shines with that sense of assurance and maternal responsibility she would bring to her later management style; the little baby boy wails with his birth-order injustice of being fifth in line, "left behind" but sure to make his presence known. I can hardly begin the analysis of my own image, an eleven-year-old painted by a seventeen-year-old viewed by a more-than-fifty-year-old. The gesture I am shown to make is ambiguous. Am I protecting my younger sister, or constraining her movements? Isn't my smile a bit tight, the shoulders hunched with a certain tension felt more by the seventeen-year-old than the prepubescent girl? These questions can never be answered; the artifice of a conclusion is tailored to the goals of narrative, not life. Similarly, just because my youngest sibling was not in this picture does not deny the enormous role she continues to play in my extended family. Mascot of the high-school marching band, moralizer to her nieces and nephews, and greatest appreciator of my cooking, she has never commented on her absence from this painting. Perhaps, unlike me, she registers the *other* absences first.

There are many narratives in this painting. Mine has been the absence of my retarded sister. Another might begin with the absence of the parents. The family shown is one constituted by relations among siblings who seem to experience themselves partly as a cohort and partly as a miniaturized enactment of the larger family itself. The oldest sister sits as the mother would, cradling the squalling infant with a confident smile. As for me, I seem to be posed as another kind of parent, a little father perhaps, standing as I would want that father to be, stalwart and steadfast in support of that almost frightened little girl. Just over my head is the only spot where the confected background threatens to open out—a small gap emerging behind and above me, even as my gaze seems focused on something outside the frame. These gazes (and that opening) threaten to split the painting. My gaze and the youngest girl's seemingly converge, looking outside, to the left and above the frame. The eyes of the standing brother and oldest sister share a different focus, more socialized and ingratiating, given to something within the frame—an implied viewer constituted, in photographic practices, by the camera's lens, or possibly by the photographer. Their gazes meet the viewer's, giving the picture its point of view and whatever comedic value it might possess. The baby is not given a gaze, as if he has no interior within that countenance—or perhaps he is *all* interior, all the time. Infant despair folds his features as if unmediated by structures of consciousness, forming a wrinkled yet continuous surface in which any interior is always instantly exteriorized, neither censored nor composed. His expression approaches comedy only in this context—since we convince ourselves that his pressing needs are purely animal, simple (put me down, feed me, change me, don't make me sit still).

The bodies of these children are rendered in such a way that they are more like a map of adjacent territories than interacting three-dimensional forms (skin touches skin in only two spots—my fingertips on my sister's arms, and the oldest sister's arm under one of the baby's). Again, there are technical reasons for this, such as the anatomical ineptitude of the student painter (defeating the artist's scant ability to depict structures under surfaces of cloth or skin), or the conversion of black-and-white photographs to color (without the necessary skills to compensate by rendering the secondary hues reflected back onto adjacent colors). But to leave you with these technical explanations would, again, be insufficient. Their partiality would foreclose the range of other interpretations that might glimmer between then and now, between student artist and art historian, between the person reflecting on her childhood and the object that speaks to her from a mysterious past. Interpretation always belongs to its present, yet mere technical explanations are never adequate to the lived complexities of the past.

None of what I have just said is known to the 1966 child or her 1972 limner. These narratives are drawn out of the painting only after decades, pulled out of this composition by the later version of the person situated within it, at a moment when there is sufficient distance to put the blended strains together to form a story for the present. It is not so much true, as a truth—a truth of evocation, not locked in this configuration, but elicited from these pigmented surfaces by present interest and desire.

In the end, this object's evocations move beyond autobiography to philosophy. When we look and think about a painting, we are making a thing coherent and meaningful, and as it accrues meaning its very objectness becomes unstable. Such instability is crucial. It is important that these are not real people, otherwise we could not stare at them, hang them on the wall, and so forth. It is equally important to the maker and viewer that they are *people,* imaginatively given inner lives, feelings, motivations, historical fates. In this respect it is intriguing to think of such a liminal thing in terms of D. W. Winnicott's theorizing of the transitional object, an important object (a teddy bear, for example) that the child experiences as both part of himself and not-himself: "The transitional object is never under magical control like the internal object, nor is it outside control as the real [primary parent] is."[1] For Winnicott, each departure into independence, each transition into separate existence, could be mediated by this evocative object.

The Winnicottian object is one whose inertness is as important as its aliveness. The teddy bear can be violently attacked, aggressively loved, fully animated, or completely ignored. The child knows it is inert and yet willingly animates it for imaginative purposes. The transitional object is needed periodically to stabilize the maturing individual—bridging between the infant's illusion of total omnipotence and the disenchanted world he must enter as an adult. Winnicott's ethical position in relation to the holders of transitional objects provides a parallel to my thinking about the "object relations" pertaining between viewers and paintings—even when, as in this case, the viewer and painter might be the same person dispersed over time. Winnicott puts it this way: "Of the transitional object it can be said that it is a matter of agreement between us and the baby that we will never ask the question: 'Did you conceive of this or was it presented to you from without?' . . . The question is not to be formulated."[2] Did I make this painting? Or was this painting made from external circumstances, "presented to [me] from without"? The question is not to be formulated, for any answer would be beside the point of evocative objects.

As we evoke meanings from the special objects we call art, we become their willing subjects. We think with them, in order to think ourselves into coherent subjectivity. We presume a homologous relation between author and viewer, but as I have argued here, some objects remind us that there can be no such equivalence. Even the author is not equivalent to her later self. It is the point of artworks to be evocative objects, soliciting us to be their subjects, and, in turn, the author of their meanings, at least for a while.

Caroline A. Jones is Associate Professor of the History of Art at MIT.

Thus the shadow of the object fell upon the ego, and the latter could henceforth be judged by a special agency, as though it were an object, the forsaken object. In this way an object-loss was transformed into an ego loss and the conflict between the ego and the loved person into a cleavage between the critical activity of the ego and the ego as altered by identification.

—Sigmund Freud, "Mourning and Melancholia"

THE SUITCASE

Olivia Dasté

It is here, still closed, in front of me. This is my grandmother's suitcase, one she would have used when she came from France. It is small, just large enough for one person to pack for a one- or two-week trip. It is firehouse red. It has tan leather handles and an exterior made of a sturdy canvas material. It has two large golden buckles, several zippers, and interior compartments. It is the perfect balance between elegant and practical, just like her. The logo boasts "Globe Trotter," echoing my grandmother's love of travel. With her newfound liberty after her husband and children had gone, she began to discover the world adventuring to Egypt, Sweden, and several times to America. But this suitcase is new; she had been saving it for one final trip.

We stayed in her apartment in Bordeaux only for a few nights, just long enough to attend the funeral and embrace loved ones. It was the day after she died, her body not even cold, when my mother began furtively emptying out my grandmother's apartment. My father and I, exhausted by grief, walked around her apartment, fearful to displace anything, looking for her. We had been there just a few weeks ago; she was doing wonderfully.

My mother faced the departure of her mother-in-law full frontal and leapt into a mechanical frenzy of sorting, organizing, and throwing away. With each book, shoe, and coat my mother grabbed and threw in a trash bag for donations or garbage, my stomach turned and my heart sank. The evidence of her life was being erased. In the kitchen, I held the glass she used the day before, fresh lipstick marks still on the rim. She would be back. She is still here.

There was no stopping my mother's rampage: my father was too drained to confront another problem; my aunts and uncles were standing like sheep before a hurricane; and my brother was too busy trying to comfort me to deal with my mother. Putting down the glass, I determined to confront her, pausing only when I saw the suitcase, on the upper shelf of my grandmother's closet, shining like a sign.

Hiding in my grandmother's bedroom, I placed the suitcase on the bedcover and opened it. The emptiness I found echoed my own. I had to act fast. I closed my eyes and pictured my grandmother as I had seen her last. She was wearing her favorite navy skirt, formed to her curves by time, the light blue collared shirt with the dancing butterflies, her well-loved red cardigan, and her white necklace. With movements that seemed automatic, I went to collect these and other items she would need.

The red cardigan still had her scented handkerchief folded in the pocket; her white plastic pearl necklace was scented with her perfume and still had her foundation rubbed in. From the kitchen I took our two pink-and-green flower-painted teacups in which we had our morning coffee. I was taking our breakfasts with us; our long, animated conversations; our ritual of sharing our dreams; our nightmares and laughter, often aimed at the least sane in our family, the two of us often being the likeliest candidates. I carefully packed the crystal butterfly my brother and I had given her for Christmas, reliving the memory of picking it out for her, together saving for it, excitedly giving it to my father to take to her in France, and later her gentle teasing over the phone, scolding, "You must be crazy! What have I done to deserve this? My granddaughter is absolutely nuts!"

I packed the cut-out quotes about love, family, humor, and life that she kept everywhere around her apartment, on the walls, in books, in drawers, and silently promised her that I would live by them. I added her jasmine-scented face lotions, the ones that I would kiss on her face each night, and her black hairbrush with her silver hairs, seeing her fuss as she did each morning until she found her *mise-en-plis* acceptable enough to go out. And I added the pictures and letters. My grandmother, I discovered, had kept each and every letter I had written to her, from the very first one, written by my pregnant mother, as me, to my last one, received three days before. She had kept my scribbles, jokes, stories, drawings, stickers, poems, and all of the pictures of me and my brother growing up. The suitcase is for both of us. It holds her for me and me for her. At the chapel, I reluctantly contemplated the body I no longer recognized. It was at the airport, hugging the suitcase, that I felt her heart. I would not let it go until it had to be stowed away in the overhead compartment. I found myself wishing I had bought an extra ticket, an extra seat for the suitcase next to me.

A year later, I have not opened the suitcase, but today I have slipped inside her letters to me and some pictures of us together. I fall asleep with the suitcase in my arms, but increasingly, it feels dangerous to open it. Memories evolve with you, through you. Objects don't have this fluidity; I fear that the contents of the suitcase might betray my grandmother.

Two and a half years after I packed the suitcase, I begin to open its buckles, one at a time. Unable to go further, I leave it like that for a while. Only now I place it on my bed and slowly begin again, determined this time to open it but working so much against myself. Finally, I have only to lift the top, but I am not ready for the smell of her perfume, her hair, jewelry, and clothes to come at me so fast. She reaches me from inside. I close my eyes, my face already wet as I find her red sweater. I don't have to pull it to my face to feel as though I am hugging her tight, but I do. I smile. I am with her in Bordeaux and we have all the time in the world.

The suitcase brings her back to me with the worry that I will lose her if I open the suitcase too often; her smell will evaporate, the letters will fade, and the clothes will no longer hold her shape. I think she would tell me to live with the living and to be careful: craziness runs in the family.

Olivia Dasté worked on the Research Staff of the MIT Initiative on Technology and Self and now lives in Paris.

Objects of Meditation
and New Vision

Bold, overhanging, and, as it were, threatening rocks, thunderclouds piled up the vault of heaven, borne along with flashes and peals, volcanoes in all their violence of destruction, hurricanes leaving desolation in their track, the boundless ocean rising with rebellious force, the high waterfall of some mighty river, and the like, make our power of resistance of trifling moment in comparison with their might. But provided our own position is secure, their aspect is all the more attractive for its fearfulness; and we readily call these objects sublime, because they raise the forces of the soul above the height of vulgar commonplace, and discover within us a power of resistance of quite another kind, which gives us courage to be able to measure ourselves against the seeming omnipotence of nature. . . .

But with this we also found in our rational faculty another non-sensuous standard, one which has that infinity itself under it as a unit, and in comparison with which everything in nature is small, and so found in our minds a pre-eminence over nature even in its immeasurability.

—Immanuel Kant, *The Critique of Judgment*

CHINESE SCHOLARS' ROCKS

Nancy Rosenblum

How can a rock, the quintessential physical object, be metaphysical? How can a stone sing? Where does nature stop and culture begin?

Tap a stone and it rings, as if it were a cast metal bell. It is resonant. A black *Lingbi* rock has an astonishing, glossy skin. A *Taihu* tilts dizzily to the side as if it were an overhanging peak, embodying what the Chinese call "awkwardness." A *Ying* has veining and a wildly wrinkled surface, suggesting age. My rocks are un-rock-like. They are plain limestone contradicting itself. The most earthy and banal material transcends itself to become exotic.

Gaze at a stone and it disorients. Scholars' rocks can be tiny miniatures an inch tall or dramatic free-standing "mountains" thirty feet tall. The same shapes and turnings, contours and depths are replicated in every dimension. The rocks draw me into the mystery of scale. A tiny scholars' rock grows before my eyes. It becomes a mountain. It is huge and theatrical. It is a whole landscape, a whole world. Its holes of many sizes and directions, its wild punctured surfaces, give it infinite depth. It contains deep space. There is movement inside. Its meaning comes not just from its contour but from the forms within. Looking at the holes is like looking at the stars. It is a world within a world.

The rocks' disorienting effects are specific and distinctive. One is the deliberate confusion of scale. Another is material turned immaterial. Another is infinite, immeasurable depth and movement in a finite space. The result of looking steadily is a direct confrontation with cosmology: how big is this rock, this earth, this universe? Blake's "world in a grain of sand," a commonplace now, is made uncommon and vivid.[1] A rock is "a little piece of a wrinkle from which you can imagine the whole wrinkle; it's a little piece of a rock from which you can imagine the whole rock; it's a little piece of a mountain from which you can imagine the whole mountain—and so on."[2]

I am not of metaphysical temperament. My rocks are compelling, though. They have the power to provide an effortless, aesthetic experience of mystery. Of infinity in a finite space. Of transformation. Just by looking. Without philosophy. Without hard ideas (a relief for an academic). They are simple, immediate invitations to playful speculation and to the wonderful physical play of handling these objects, turning and inverting them to find new perspectives.

Scholars' rocks are mounted on beautifully carved wooden stands. As early as the Neolithic period in China, ritual objects were placed on pedestals to lift them into the realm of the sacred. Over time their religious meaning was eclipsed, and the stand turns the rock into an art object. The stand is part of the rock's transformative character, for removed from its stand it flips back from art to nature. The stands are more than the pedestals for display that accompany all sorts of decorative objects, then. The stands add an iconographic element, too. They can be read. The bumps, or "teats," on stands for stalactite rocks evoke the mountain teats through which the milk of the earth flowed. (An early Taoist theme has it that somewhere in the highest mountains is a cave that is an exact representation of the world outside. In its center is a stalactite that gives off the milk of contentment.) Stands carved with clouds evoke another transformative idea, that rocks are petrified "cloud roots." Stands carved with water and waves enhance the *taihu* rocks, with their infinitude of holes made and multiplied by immersing the stones in moving water—microcosms of the earth's geology. The stand's stylizations are a clue to the age and region of the object when it was first taken from nature.

Setting a rock on a stand has another effect. It isolates a piece of nature, removing it from its original context. It points up the resonance of a single natural object, bringing home individuality, a thing complete in itself. It points up the inexhaustibility and beauty of randomness and irregularity, clearly. And because the rock flips back and forth between nature and culture, it brings home transformation. Again, difficult concepts are made simple and immediate. This is nothing like the isolation of everyday manufactured objects—urinals or debris from a scrap yard—by contemporary conceptual artists.

Scholars' rocks are the point at which wild nature and culture meet. Many of them are "worked" in some way, but always so that the hand is hidden. Only high-power glasses reveal the tiny chisel marks that enhance the hollows and crags. They are what contemporary expressionism calls controlled accidents. The invisible sculptural element of the scholars' rock is integral to it as a transformative object: inanimate to animate, nature to culture.

The distinctive thing about the rocks is that the transformation is reversible. The rock flips between nature and culture. Renaissance sculptors envisioned a figure in a marble block, but for them nature was just the starting point. The ideal, as in Michelangelo's *David,* is to leave nature behind. The rocks, even when they are representational, are not denatured. They preserve duality.

Scholars' rocks raise conceptual issues of authenticity, authorship, and what it means to make something. But these academic questions are not part of their resonance. The cultural arcana of the rocks are not part of their resonance either. These questions are ancillary to the power of the rocks to flip in the ways I have described, and to likewise flip me from mundane to mystery.

Chinese rocks have a long history in art and religion. By the Song Dynasty (960–1279) the rocks became prized by the literati for whom they were objects of contemplation. They brought wild nature into the studio. The rocks gave shape to the strange design of classical Chinese gardens, which echo not green nature but the pools and twists and turns of the inside of caves. They were the subject of poetry. They were models for the paintings of rocks and mountains that dominate Chinese scroll painting. The rocks were treasured by emperors, and Chinese literature is full of stories about the quest for these revered objects. "Scholars' rocks" come from their work, but the rocks go deeper into the past of early Taoism, most likely as early as the second century BCE when they were first excavated from lakes and underground caves.

Chinese scholars' rocks are chosen, worked, mounted because they are evocative of mountains, the home of the gods. Their power to evoke the experiences I have described are an important part of the history of Chinese art and culture. The experience of these rocks is not subjective. They *are* evocative objects. They are gifts of nature and Chinese culture designed to be evocative; my experience with them is not my own alone.

And yet these stones do have a purely personal, emotional aspect for me. My late husband, Richard Rosenblum, brought these rocks back from neglect in China and oblivion in the West. He trusted his own way of seeing. The art history profession and the museum world, proud of its universal inclusiveness, had missed a major art form from a refined culture.

The rocks made our lives together an adventure. Richard and our daughter Anna looked for them in unlikely shops and botanical gardens and the homes of individuals in London, Hong Kong, Taiwan, and many parts of China. The rocks filled our home—every floor and every surface was covered with miniatures and giants. Richard gathered black and gray and green *lingbi* and pure white *taihu,* marble dream stones with images in their veining, and "Chrysanthemum stones" with their gorgeous fossils. Rocks of every mineral type, from soapstone to pudding stone, and "imitation" rocks in crystal and wood, ivory, glass, bronze, and sophisticated glazed ceramic.

Richard brought together the entire canon of prized formal elements. He organized and studied them and commissioned academic studies of them—technical scientific studies and art history essays to begin to create a body of scholarship. He organized shows and exhibited the rocks around the world; he wanted them to be known, especially to artists. He and our daughter donated them as gifts to museums. Together we wrote essays and a book. Richard was an artist, and his own sculpture, made from natural tree roots and branches, flipping between nature and human image, preceded his discovery of the rocks. The rocks added meaning and inspiration to his work. They excited his sculptor's interest in scale and his unique interest in constructing deep space in both sculpture and digital prints.[3]

How can a rock be a man? "The Honorable Old Man" rock is Richard for me—obsession, looking, openness to being surprised and moved, dignity.

Nancy Rosenblum is Chair of the Department of Government and Senator Joseph S. Clark Professor of Ethics in Politics and Government at Harvard University.

We are told that the trouble with Modern Man is that he has been trying to detach himself from nature. He sits in the topmost tiers of polymer, glass and steel, dangling his pulsing legs, surveying at a distance the writhing life of the planet. . . . Nor is it a new thing for man to invent an existence that he imagines to be above the rest of life; this has been his most consistent intellectual exertion down the millennia. As illusion, it has never worked out to his satisfaction in the past, any more than it does today. Man is embedded in nature.

The biologic science of recent years has been making this a more urgent fact of life. The new, hard problem will be to cope with the dawning, intensifying realization of just how interlocked we are. The old, clung-to notions most of us have held about our special lordship are being deeply undermined.

And nothing would better describe what this place is like, to an outsider, than the Cézanne demonstrations that an apple is really part fruit, part earth.

—Lewis Thomas, *Lives of a Cell*

APPLES

Susannah Mandel

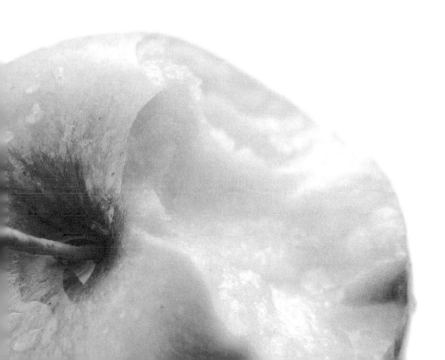

As far back as I can remember, I have had an unusual fondness for apples. When my roommates offer to pick up fruit at the grocery, they are often startled by my specificity: "Galas are festive, but only if they're brightly streaked, red and yellow like a fall leaf. If the skin seems dull, buy Fujis—you'll see them stacked in a chilled mountain—or some Pink Ladies. But make sure they're ripe; the proper color of a Lady is like champagne, or the rose they used to tint women's cheeks with in old photographs. If they've gone too soft you can get regional apples—you'll know them by the smell of local orchards—McIntosh, maybe, but only if they're sharp red and white. If they still have a cast of green, go for Jonagolds, or Winesaps, or Empires from New York. . . ."

"I thought you only ate that apple from New Zealand," the roommate might say at this point, standing in the doorway and giving me the fishy eye.

"That was last year. New Zealand apples are magnificent, but this season the Braeburns have gone wine-y."

This is more than connoisseurship. I have a *thing* for apples. For years, the apple has been my favorite aesthetic object—as well as a really good element of my lunch. Since childhood, I have filled my pockets with apples and gone around with my coat distended like some sideways marsupial. I put apples on my nightstand, as some people place candles. I learn their names. And, of course, I also put them to their best and natural use: plucking them out of the places they have been tucked away (usually the pockets), and, at home, on city streets, in the subway, or in forbidden corners of libraries or museums, biting into them. If you close your eyes, as you would in a kiss, it helps you relish at once the cool solid weight of the ripe fruit in your hand, and the smell and sound and taste of the smooth skin breaking, and all the sweet juices leaking out.

As a child, one of the things I liked best about apples was their aura of stolid historical continuity. Each apple is unique in its excellence. But every apple is also trivial, disposable, and identical to the one I ate this morning and the one I will half-finish tomorrow. The apple currently bulging my pocket has infinite echoes, infinite siblings, suggesting a gallery stretching back in time—an endless banquet of fruits. Consider: if my apple is the same apple as yesterday's and tomorrow's, then it follows that it is also the same apple as the apple of two or four hundred years ago. Shakespeare ate my apple! And the Romans when they first came to the gloomy, rainy island of England. It feels luscious to know you are eating something historical; it makes you feel rooted in time.

On our bookshelves when I was young, my parents had a dog-eared copy of Howard Pyle's *The Merry Adventures of Robin Hood,* dated from 1883, as well as lots of other nineteenth-century medieval adventures and Lang's Fairy Books. In these jolly stories, people were always quaffing ale, roasting venison, and journeying through forests with cheese and apples in their packs. I liked to emulate their dark-age lunches, reading on the living-room carpet, with cheddar cheese and the end of a loaf of bread. I enjoyed the feeling—eating as I read— that, as I crunched through an apple's skin and juice, I was having precisely the same experience as that of Robin or those youngest sons of tailors as they journeyed through the wood. Fashions may change, but the tang of an apple in an American living room is, must necessarily be, the same as the tang of an apple in 1194. Such small questions as the changes wrought by fruit husbandry over the years must yield in the face of such transcendent eating experience. The sensory experience of the bite collapsed the centuries together, fusing myth and reality. In the timeless now of imagination, Robin and I sat against trees or bookshelves and ate our lunch together, the juice running down our chins, both in the same moment.

My relationship with the apple has changed with the years. Until about the age of twelve, for example, I preferred green apples. My parents were fans of the health benefits, and offered them frequently as both lunch staple and dessert. Green apples have the homely name of "Granny Smith"; you feel you could be on speaking terms with this fruit. There's something in the firm solidity of a Granny Smith that makes it feel reliable— it is just slightly larger than a child's fist, something the hand can curve around, with an understated tactile pleasure, like a toy ball.

At the same time, the apple's glossiness of skin— something in the smoothness of the texture—makes each green Granny seem to be like every other. One gets the idea they could come off assembly lines, like the plastic fruit that comes with children's toy kitchens. They are indistinguishably sweet, infinitely repeatable, and perpetually green.

At twelve, my taste changed abruptly. I had loved green apples; suddenly, now, I would only eat the kind with red-and-gold stripes. It paralleled a change in my fondness for flowers—a liking for pink or yellow roses and tulips changed into a passion for the kind that's striped like a tiger.

My mother threw up her hands: how arbitrary are the loves of children! But there was aesthetic meaning here, as certainly as there is meaning when one's tastes shift in literature or music. A solid pink flower petal, like a green apple, is intense and sweet. That's where its power lies. In the denseness of its flesh and its color, it inspires thoughts of happy things, like perhaps an

Easter dress or the sun. It flames up in the gaze, but it doesn't hold it, and the eye, momentarily arrested by the fierceness of the blaze, soon disengages and slips over it and away.

A striped fruit or flower, on the other hand, will catch your gaze and keep it. It will not let you look away. When I held a red-and-yellow Gala or New Zealand Braeburn in sunlight, I saw with fascination how it seemed to catch fire from inside. I turned it over and over, looking, and always seeing something new. Apples' stripes are labyrinthine. Their variations, their variability, changed the aesthetic of the "pretty" and the "sweet" into something deeper, a reminder that nature is complex and unpredictable. Each is as unique as a snowflake, or a big thumbprint. You can gaze into such patterns endlessly.

Later I learned about the meditative practices of medieval monks, who would spend long sessions contemplating labyrinths, as monks in Asia today still contemplate mandalas. They were trying to open their minds to the infinite. The pattern was a little image of contained eternity, small enough to fit in a temple or be engraved on a cathedral floor. But it stood in for everything in the world.

The link between labyrinths and apples is evoked in a story from Julian of Norwich, a fourteenth-century English mystic who claimed to regularly talk to God. She wrote that one day God showed her something like a hazelnut lying in the palm of her hand. When she asked, reasonably enough, "What might this be?" she was told: "It is all that is made." I am pretty far from being a fourteenth-century mystic, but I find something warming in the idea that all the world's beauty and oddness can be symbolized in the complex patterns of something small enough to fit in your hand.

From where do we take our lessons about simplicity and complexity? At some point in my youth, a wooden apple entered my life. I don't remember when it appeared, or where it went; it came from somewhere and remained on the shelf of the room I shared with my brother, disappearing out of our lives again several years later, as if evaporating into smoke. It was the size of a real apple, surprisingly light—perhaps it was hollow—and painted to resemble some red-and-yellow strain, of the Braeburn or Fuji type, with a varnish that gave it a dim but realistic glow. A paper leaf drooped from an abbreviated wooden stem.

As a singular phenomenon this apple was interesting, and I thought at first that I found it beautiful. I should have wanted to hold it, play with it, and keep it in my pockets. But I realized with a slight sense of guilt, that it was actually very boring. Its lightness, its gloss, and perhaps the fact that it was never going to be eaten and reduced to seeds and core, left it lacking.

I was certainly not the first to notice the strangeness of an artificial fruit. In *A Clockwork Orange,* Anthony Burgess uses strong metaphors to evoke the unnaturalness of a world that removes from its citizens the possibility of moral choice: "The attempt to impose upon man, a creature of growth and capable of sweetness," writes one character, "to ooze juicily at the last round the bearded lips of God, laws and conditions appropriate to a mechanical creation, against this I raise my sword-pen."[1] As Burgess records, American readers, who didn't realize that "clockwork orange" was old Cockney slang used to describe "anything queer," took from it the secondary image of "an organic entity, full of juice and sweetness and agreeable odour, being turned into an automaton."[2]

There is something powerful in this image of a fruit made into a machine—something striking in the unease it engenders. But this negative view is, perhaps, unfair to the wooden apple. The artificial fruit on the shelf is a wonder in itself, although, for those who are looking for signs of life, for reassurance that the universe is still alive and kicking, it is far better to be able to touch the real thing.

The questions that we start asking in childhood stay with us, growing as our minds grow. Today, when I look at modern art, I find myself still searching for those who play with the tension between the natural and the artificial. René Magritte, whom I love, liked to paint both apples and an infinite number of men in bowler hats. He walked the line that divides the natural from the human-made, acknowledging a line that is increasingly blurred. Magritte's repetition of suits and hats, umbrellas and pipes, also mixes in obscure symbols of "the natural world." Sometimes he renders identical businessmen, as thoroughly multiplied as androids or apples, falling like gusts of rain from the sky; sometimes the businessman stares out at us, face hidden behind a mysteriously suspended green apple. Magritte's apples are not particularly unusual in appearance—they are perfect and round like all other Granny Smith apples. But something else about Magritte is that he also liked to insert unexpected doors in his scenes: a hole ripped in a wooden door, a smashed picture window. It may be that, in this identical world of suits and suburbs, the green apple should also be seen as a gate into a different and refreshed world of aesthetics and philosophy, as a safeguard against the slightly weary moment approaching in which the most dedicated fan of human-made objects might get tired of all those bowler hats.

Why, in the end, do I carry an apple in my pocket? To put it one way: the apple is a good talisman because it can stand both as a symbol for nature's careless sprawl, and as a focal point for the intense emotion, or contemplation, we sometimes need these symbols to evoke. But the apple has another, final strength, which is simply this: it resists being absorbed too far into symbolism and the sublime. It's too close to the ground; it's still got dirt stuck on it. On this front I am reminded of a suggestion made by the physician and essayist Lewis

Thomas in *Lives of a Cell.* Thomas writes that when we go beaming information into space, to let any hypothetical aliens out there know what we're like, we must be sure to include some of Paul Cézanne's still-life "demonstrations" that an apple "is part fruit, part earth." "Nothing," Thomas adds, "would better describe what this place is like."[3] By which, of course, he means life on Earth.

Life on Earth! Dusty, passionate, sublime. Or maybe sometimes, an apple is simply an apple: rich in our mouths, heavy in our hands, sweet in our pockets.

Susannah Mandel is a recent graduate of the MIT Program in Comparative Media Studies, where she studied comic books.

The "double" was originally an insurance against the destruction of the ego, an "energetic denial of the power of death" . . . ; and probably the "immortal" soul was the first "double" of the body. This invention of doubling as a preservation against extinction has its counterpart in the language of dreams, which is fond of representing castration by a doubling or multiplication of a genital symbol. The same desire led the Ancient Egyptians to develop the art of making images of the dead in lasting materials. Such ideas, however, have sprung from the soil of unbounded self-love, from the primary narcissism which dominates the mind of the child and of primitive man. But when this stage has been surmounted, the "double" reverses its aspect. From having been an assurance of immortality, it becomes the uncanny harbinger of death.

—Sigmund Freud, "The Uncanny"

THE MUMMY

Jeffrey Mifflin

I have in my custody as the archivist and curator at Boston's Massachusetts General Hospital a large and significant collection of historical objects, including paintings, chandeliers, medical and surgical instruments, antique baby bottles, and a horse-drawn ambulance. The most unusual object in the MGH's historical collection, however, is undoubtedly its 2,650-year-old Egyptian mummy.

On an August day in 1998, before my initial job interview, I explored the circuitous halls of the hospital, making a pilgrimage to the famous Ether Dome, the old operating amphitheater, where in 1846 the first public demonstration of surgical anesthesia had taken place. I was astonished to find there an Egyptian mummy in an open coffin brightly decorated with painted hieroglyphics. Why did the hospital have a mummy? And why wasn't it being better cared for? Its exhibit case was dusty and scratched. Its leathery, exposed face was freckled with a white substance resembling the bloom on a plum. Its desiccated lips were drawn back in a toothy grimace reminiscent of the scary faces I had seen in horror movies as a child.

I now know from research in the documentary record that Dutch merchants trading in the Ottoman Empire donated the mummy to the hospital as an anatomical specimen in 1823. The mummy's name, occupation, and place of origin are known because an Egyptologist from Boston's Museum of Fine Arts translated the coffin's hieroglyphics in 1960. They read, in part: "Spoken by Osiris, he gives all food offerings for Padihershef, deceased, son of Iref-a-en-her, his mother, the lady Her-ibes-ines, deceased. Greetings to thee Osiris." Padihershef, a stonecutter who lived near Thebes during the Saite Period (XXVI Dynasty), died in his late forties. Once mummified, he was placed with other mummies in a group burial cave. In 1823 he was exhumed and stashed in the hold of an American vessel, along with barrels of raisins and cases of opium for medical use in New England, and shipped via the Mediterranean and North Atlantic to the new world. His title, *Hrtyw-ntr*, meaning "stonecutter in the necropolis," reveals that, in life, he earned his living by tunneling through the limestone cliffs on the West Bank of Thebes to make tombs.

Padihershef was one of the first Egyptian mummies in the United States. He became famous when MGH surgeon John Collins Warren unwrapped his "twenty-five thicknesses . . . of bandage . . . imbued with some glutinous substances, intended to preserve them" before an audience of scientific men and published an illustrated report, "Description of an Egyptian Mummy," in the first issue, in May 1823, of the *Boston Journal of Philosophy and the Arts*. His exposed face (which was never again covered) still tends to exude the "whitish saline efflorescence" described 181 years ago by Warren in his report.

In preparation for my first mummy loan I spent many hours negotiating and drafting terms for an agreement, locating and hiring experienced mummy movers, and wrangling over insurance coverage. As part of the negotiations surrounding my second mummy loan I secured an agreement whereby the borrowing museum would pay for conservation by a professional mummy conservator, which would include cleaning (and at least temporary elimination of the efflorescence). During these periods of research, measurement, and photography, the mummy and I became very well acquainted.

Some time ago I opened Padihershef's wood-and-glass exhibit case after it had been tightly sealed for several years. When I put down my tools and lifted off the front panel of the case a pungent aroma of spices and resin suffused the room, a sensation common, perhaps, for a 650 BCE Theban nose, but unforgettably peculiar for an American in the twenty-first century. (Dr. Warren was similarly surprised by the same smell in 1823.) I realized that the sensation stimulating my nostrils at that moment was the same experienced by Padihershef's friends in Thebes just before they lowered the lid on his coffin centuries before. I looked at the mummy in a new light. He was less abstract, and more like me. He had been flesh and blood and bone, and the flesh and bone were still there. His senses had once worked as mine now did. His mind was gone, but neither would I live forever.

I had crossed paths with other mummies as a child and as a college student, but had never known them in the way that I came to connect with Padihershef.

As a teenager in St. Louis I had wanted to get away and immerse myself in interesting work in a distant place. The rooms in the St. Louis Art Museum that drew me like filings to a magnet were those containing Egyptian mummies. The mummies symbolized the mysteries of the past, the enigma of time, the unfathomable depths of the unknown. But they also helped me, I think, to bridge the gap between the knowable and the inconceivable. I wrote to archaeologists asking about summer jobs, hoping for the chance to sift through and touch and photograph the evidence of the past. I got back several polite replies on university letterheads thanking me for my interest and suggesting further study in preparation for a career. Such ambitions were considered impractical by my mother and father and were strongly discouraged.

In the late 1960s I enrolled as a student at the University of Chicago. In 1968 I remained in Hyde Park for the summer quarter, working as a night janitor at the Oriental Institute Museum. A mimeographed schedule of duties placed me in the mummy room around midnight each Monday through Friday. The room contained

five Egyptian mummies in glass exhibit cases, surrounding a giant statue of the boy king Tutankhamen,
with smaller cases displaying canopic jars, amulets,
and mummified ibexes. My duties included washing the
glass and turning off the lights in the museum, as well
as in the jumbled storeroom in the basement, surrounded by human remains and shifting shapes in the
half-light. Museums at night are dark and echoing
places, and late adolescence is a troubling journey. My
time there was unsettling.

I had recurring thoughts as I perfunctorily did my
janitorial work. The mummies in the exhibit cases, I
supposed, had believed they would live forever. They
had saved enough in life to begin their death wrapped in
several hundred yards of linen. Religion (which I no
longer had) was a comfort to those who believed. The
scholars who wrote the exhibit labels in the museum
were, for the most part, transmitting accepted assumptions and guesses. The chaos in storage, which the public never saw, but which I experienced at close range,
gave an entirely different impression from the orderly
presentation of supposed facts carefully packaged for
general consumption in the public exhibits. Would I ever
be able to secure such a degree of certainty about things
that mattered?

Padihershef's life is over twenty-six centuries and
thousands of miles removed from mine. Was life simpler
(or better) for Padihershef than for me? Did he enjoy love
and the satisfaction of accomplishment? I hope so. Did
he also know fear, pain, disappointment, and failure,
those seemingly inevitable companions of the human
condition? Did he have historical perspective, or suffer
from existential angst?

He knew a few things very well and probably believed in the ordered afterlife described by the hieroglyphics on his coffin. Surrounded as he was in Thebes
by tombs, ruins, and inscribed obelisks, he probably
had some awareness of the historical past. How did the

reverberation of his bronze chisel on stone, or the rough texture of his wooden mallet, or the smell of a burial evoke for him the sense of his personal past, and how did this affect him? What was his personal measure of himself and how did it compare with the expectations of his society?

I have calculated the generations back to ancient Egypt. There are about 130, I estimate, between Padihershef and me. I was born in 1949, my father in 1919, and his father in 1874. My ancestors came to America from Wiltshire, England, in the 1680s. Grandpa, Great-Grandpa, and the rest line up in my mind like dominoes poised to fall, or megaliths waiting to be worn away or covered up by centuries of weather and accumulated strata. Was my distant progenitor in Britain mining tin or slicing blocks of peat at the same time that Padihershef was chiseling out tombs in Egypt? What was my place in this family lineage, this historical continuum? I have, in due course, occupied my allotted place and represented my generation. My digging and tunneling have been of a different sort, reading and writing, exploring, analyzing, and making available the historical record.

There is an inner frontier as complex and elusive as any external exploration ever attempted. My wonder, I expect, about my own place in the continuum will continue for as long as I live. As Pascal remarked, "When I consider the short duration of my life, engulfed in the infinite immensity of spaces of which I am ignorant . . . I . . . am astonished at being here rather than there."[1] I still wonder at circumstance, chance, and the lineage of things that matter.

Jeffrey Mifflin, an archivist who specializes in technology, science, and medicine, divides his time between the Massachusetts General Hospital and MIT.

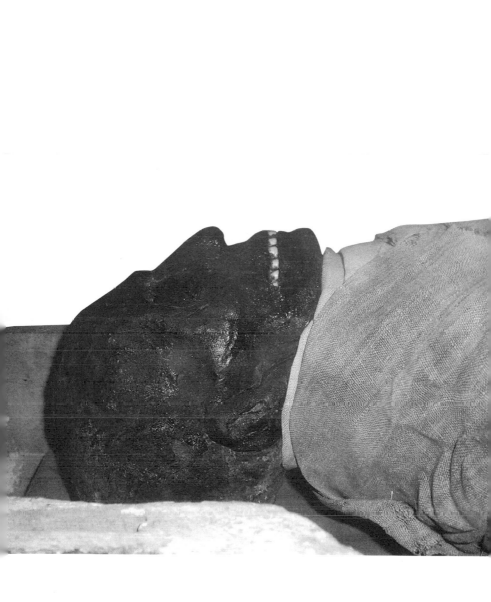

Here, and on this sentence that was perhaps also meant for him, he was obliged to stop. It was practically while listening to her speak that he had written these notes. He still heard her voice as he wrote. He showed them to her. She did not want to read. She read only a few passages, which she did because he gently asked her to. "Who is speaking?" she said. "Who, then, is speaking?" She sensed an error that she could not put her finger on. "Erase whatever doesn't seem right to you." But she could not erase anything, either. She sadly threw down all the pages. She had the impression that although he had assured her that he would believe her implicitly, he did not believe her enough, with the force that would have rendered the truth present. "And now you have taken something away from me that I no longer have and that you do not even have." Weren't there any words that she accepted more willingly? Any that diverged less from what she was thinking? But everything before her eyes was spinning: she had lost the center from which the events had radiated and that she had held onto so firmly until now. She said, perhaps in order to save something, perhaps because the first words say everything, that the first paragraph seemed to her to be the most faithful and so did the second somewhat, especially at the end. . . .

He did not remember questioning her, but that was no excuse; he had questioned her in a more urgent manner by his silence, his waiting, and the signs he had made to her. He had induced her to say the truth too openly, a truth that was direct, disarmed, irrevocable.

—Maurice Blanchot, *Awaiting Oblivion*

THE GEOID

Michael M. J. Fischer

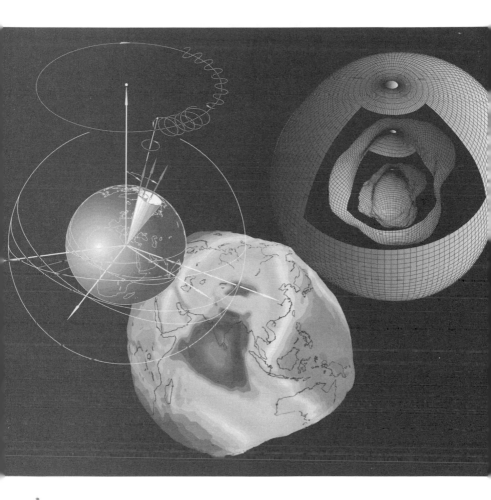

"It's *my* article!" Irene complains about her 124th publication, the first of a nine-part abridgment of her scientific autobiography, appearing as she nears her 97th birthday: "Why can't they use the title the way I wrote it?"[1] She is right, of course. She usually is, even if the words don't always come out right. My mother is confined to chair and wheelchair. Her eyes tire too quickly to read very much.

The title of her autobiography is *Geodesy? What's That? My Personal Involvement in the Age-Old Quest for the Size and Shape of the Earth, With a Running Commentary on Life in a Government Research Office.*[2] The publisher has suggested another, shorter title, *Geodesy? What's That?* and it is hardly the same. It makes the book sound like a primer, a technical introduction rather than what it is: a lively, sometimes sarcastic, and above all joyous account of a career she loved. My mother's scientific career from 1952–1977 coincided with the golden age of classical geodesy, and she reveled in her ability to participate and lead.

The geoid, moiré-like, simultaneously material earth and mathematical shape, is defined as that equipotential surface of the earth's gravity field that most closely approximates mean sea level. It is an uneven surface with flattening at the poles and bulging at the equator. There is more flattening at one pole than the other, and it is marked by depressions here and there, as around the Hudson Bay, caused by the Ice Age (like a thumb pressing in on a rubber ball). There are differences too between the sea and land, and between mountains and other geomorphological features. The Pacific and Atlantic sea levels are not the same. The irregularities of the geoid are measured against reference ellipsoids, known by the numbers defining their flattenings.

For me, the geoid begins with the high school geometry textbook my mother wrote while I myself was in high school, struggling not to besmirch the whiz-kid reputation my older sister had established with the math teachers. By the time I was learning geometry, my mother was instilling in me the basics of geodesy.

In 1952, when I was in first grade, my mother was hired to work in the four-person Long Line Section of the Geoid Branch of the Army Map Service. It was the projective geometry that Irene had learned in Vienna that began to make her name within the Geoid Branch. During World War II, two-dimensional survey grids were rotated until they fit into neighboring surveys when stretched to scale. My mother insisted that the future of map making would require three-dimensional approaches and began building the databases that would become the world datums that carry her name.[3]

As Irene rose in the ranks, eventually becoming the Geoid Branch Chief, she began to explore a series of historical problems such as why conventional geodesists insisted on using limited two-dimensional shortcuts. Sometimes Irene pursued these puzzles to break down barriers to further conceptual development. Sometimes she pursued them just to solve historical puzzles, as when she investigated why determination of the earth's circumference by Eratosthenes (the ancient philosopher from Cyrene) had been reduced to 180,000 stadia by Posidonius or by Ptolemy. This reduction had been a fateful underestimation of the size of the earth that allowed Columbus to persuade his patrons that the distance westward by sea to India and Cathay would be shorter than eastward overland.

My mother's pursuit of historical puzzles of this sort was constant confirmation to me of the value of learning languages, and of how much better my parents' European gymnasium education had been than that provided by my American high school. (My father had eight years of Latin and four of Greek, and my mother studied Latin in school, Greek after school, English and Hebrew, and later taught herself Russian to read technical literature, and Yiddish to translate a book about the Ukrainian village of her father's origin). But perhaps more important to me in my teenage years, preoccupied with the quest for philosophical clarification, was the example of the geoid as a model. It models the idea of manipulable geometric constructions against which a complicated reality could be compared, and of best-fitting ellipsoids, each fitting a portion of the earth better in different places. Historically, this work had involved famous expeditions such as the two sent out by the *Académie Royale de Paris* in 1735–1736 to measure arcs in Peru and Lappland, and such systematic surveys as when in the 1870s India trained pandits and lamas to secretly measure the distance from Darjeeling to Lhasa hiding strips of paper in Tibetan prayer wheels to record the paces marched.

The expeditions continued in Irene's day. She was part of delicate negotiations to get the South Africans, the Argentineans, the Brazilians, and others to contribute their national survey data to her geodetic world datums. As an enticement to the Argentineans for contributing their data, Irene arranged for them to be trained on and be able to use the Army Map Service's computers. Jealous bureaucrats prevented her from attending conventions at which her results were presented, but the South American geodesists gave her a quite unusual and official "Vote of Applause." Today oceanography, marine geodesy, satellite geodesy, and now, space geodesy have transformed the field. Irene can't keep up, and I can do so only as a lay outsider. We get *EOS, Transactions of the American Geophysical*

Union, and I read or summarize the articles of interest to her. The print is way too small for her to read.

Books on geoids rarely come in Large Print, and when they do, they often do not come with enough white around the margins and paragraphs to make reading easy. Listening to someone else read takes focus and attention. Becoming familiar with a book is a different process in old age if you can't flip around in the pages or use the index yourself, more like exploring a new territory with a cane. And if your memory is no longer quite so good, as is the case for Irene, things easily get confused. You fill in hypothetically, sometimes coming to closure about connecting the dots with false memories that seem so clearly right. So we read a book not by reading from beginning to end, but by zeroing in on projects and places, looking for the points of attachment, for the personal and social in the science.

My own career repeatedly crossed outposts of my mother's work. Everywhere I went, there were points of attachment, people, and memories that allowed Irene to travel along, as we might say today, "virtually." Now, the evocations of the geoid create a community that Irene and I still can share. They help to orient her world, reaffirm her personhood, allow her dignity amidst the indignities of old age. It is not so important that the dream world intervenes, that logic gets confused, that reading cannot be managed, that what was once intellectual challenge is now too technical. What is important is the self and its relations, the ability to feel oneself as sentient, as having accomplishments, as being recognized. Irene wrote her autobiography not as a primer, but as a sharing of memories, personal stories, and perspectives.

For my mother and me, the geoid has become a vehicle for the negotiations of old age, a surface for mutual reference. It keeps the two of us grounded and in synchrony. The geoid is hard as rock, fluid as the sea, smooth and distorted as a computational surface, a transitional object that draws me into mathematics, history, faraway places, and a community of people from Eratosthenes to Ptolemy, mathematicians to astronauts, and always already into the family romance that an immigrant geographer-historian and his mathematician-geodesist wife created.

Michael M. J. Fischer holds a joint appointment in the Program in Science, Technology, and Society and the Anthropology Program at MIT.

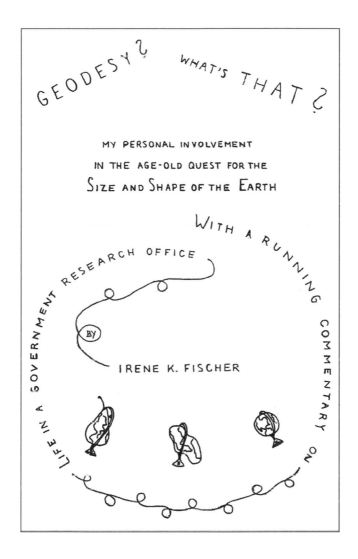

GEODESY? WHAT'S THAT?

MY PERSONAL INVOLVEMENT
IN THE AGE-OLD QUEST FOR THE
SIZE AND SHAPE OF THE EARTH

WITH A RUNNING COMMENTARY ON LIFE IN A GOVERNMENT RESEARCH OFFICE

BY

IRENE K. FISCHER

Another source of the sublime is infinity. . . . Infinity has a tendency to fill the mind with that sort of delightful horror, which is the most genuine effect and truest test of the sublime. . . .

Whenever we repeat any idea frequently, the mind, by a sort of mechanism, repeats it long after the first cause has ceased to operate. After whirling about, when we sit down, the objects about us still seem to whirl. After a long succession of noises, as the fall of waters, or the beating of forge-hammers, the hammers beat and the water roars in the imagination long after the first sounds have ceased to affect it; and they die away at last by gradations which are scarcely perceptible.

—Edmund Burke, *A Philosophical Enquiry into the Origin of Our Ideas of the Sublime and Beautiful*

FOUCAULT'S PENDULUM

Robert P. Crease

My first encounter with Foucault's pendulum left me unsettled.

I was twelve years old and visiting the Franklin Institute, a science museum in the heart of Philadelphia, the city where I was born. This was before the heyday of interactive demonstrations, but the Institute had plenty of hands-on exhibits, mostly in electricity and mechanics, that I found fascinating and gratifying. The display in the main staircase to the right of the main entrance, however, was different. It was a simple device: a heavy polished silver bob, about two feet in diameter, suspended by a wire cable from the ceiling four flights up. All the bob did was swing, slowly and ponderously, back and forth over a compass rose mounted in the floor, once every ten seconds. As the day wore on, the plane of its oscillation slowly drifted around to the left. The change was noticeable if you paid attention to the compass rose for a long enough time, but was marked more tangibly thanks to a set of four-inch steel pegs that stood in two semicircles on the floor, like a column of stout and determined toy soldiers, around the outskirts of the rose. A small stylus underneath the bob knocked down the peg-soldiers, one by one, about every twenty minutes or so. The stylus would creep toward a peg-soldier, graze it, make it wobble, and then knock it over, literally like clockwork. The sign by the first-floor stairwell said that while it may seem as though the pendulum's swing was changing direction, what the museum visitor was really seeing was the earth turning. Try as I might, I couldn't see it.

During my junior high school years the Institute became my haven. After school, I would rush there when I could—taking a train, then a bus—to have as much time as possible before making my way home. My favorite exhibit would change from month to month: a wind tunnel, a mechanical clock, an electrical display that sparked, a funnel that traced elaborate designs when filled with sand, a telescope on the building's roof. The Foucault pendulum was never my favorite, but it had a majestic presence that was unlike the others. It stood in a kind of isolation. It had no buttons to push, no dials to turn. It generated no sparks, gave off no lights, and did not hum. Institute staff members started the pendulum in the morning just before the doors opened—I never saw this happen—and then left it to itself the entire day without any additional pushes, electronic or otherwise. It was huge. It ran from the bottom of the stairwell all the way to the ceiling. It dwarfed me. It couldn't be taken in at a glance, whether you were high up in the stairwell looking down on the bob and compass rose, or at the bottom level looking at the bob and pegs dead on. I would always spend part of each visit returning to the staircase to watch its silver sphere glide silently back and forth, occasionally knocking down a peg, until some moment late in the day when the pendulum swing had narrowed sufficiently so that the bob no longer reached the outskirts of the rose and the rest of the peg-soldiers were safe. Still, stare as I might, I could never see the earth turn.

One day I went to the Institute's library and found an article about the pendulum in a back issue of the Institute's bulletin. That provided me with information about this particular pendulum, which had been one of the Institute's first exhibits when it had moved into its current building back in 1934. The article also inspired me to read up on the life of the nineteenth-century physicist Jean-Bernard-Léon Foucault, who, while playing around with a pendulum-driven clock that he wanted to attach to a camera, noticed that a pendulum continues to move in the same plane when its mount is turned, and realized that with this effect he could demonstrate the motion of the earth. Foucault's most famous demonstration—commemorated by several drawings—took place in March 1851 at the Panthéon, which used a cannonball for a bob suspended by a wire over 200 feet long. The bob glided twenty feet across the floor with each swing, making one back and forth oscillation every sixteen seconds, and at the end of each swing a stylus mounted underneath the cannonball cut a mark in a bank of wet sand.

By the end of 1851, Foucault pendulums had sprung up all over the world. Today there are hundreds, perhaps thousands: their wires are longer or shorter, their bobs made of various materials, their periods quicker or slower, and they mark the shift in the direction of the plane of oscillation in different ways. The amount of the shift depends on the latitude. At the North and South Poles, a pendulum would make a full, 360-degree circuit every twenty-four hours, moving fifteen degrees per hour. The hourly deviation is smaller closer to the equator—and the shift is clockwise in the northern hemisphere, counter-clockwise in the southern. Each and every one of these pendulums is in effect connected by the fact that they are on the same spinning globe. I found this, too, an inspiring vision. I imagined clusters of people surrounding each one of these pendulums in museums, schools, and churches all over this vast planet. They are staring at it and seeing . . . seeing what?

Armed with all my knowledge and information, I still was not sure. Could they indeed be seeing the pendulum plane as utterly motionless, and the earth—that is, themselves, the people standing next to them, the floor underneath, the building, and the entire rest of the planet to which the building was attached—as moving? Foucault himself clearly thought so, and often wrote about how his pendulum spoke "directly to the eyes." So did the museum curators who wrote the signs at most of the Foucault pendulums I have seen.

Perhaps, I thought, the pendulum is a grand optical illusion, a case of our senses presenting us with one "reality"—that two sticks are of different lengths, say, as in the Müller-Lyer illusion—while "science" (in the form of a ruler) presents us with something different. In the case of Foucault's pendulum, the collision is between our perception that our immediate surroundings are stationary and the astronomical conclusion that we are whirling about. But this would still not explain how we could see the earth moving. The ruler, after all, does not cause the Müller-Lyer illusion to cease—we still "see" one stick as bigger than the other, even though psychologists may be able to train themselves not to do so. The museum's signage does not tell me what I see. My vision does.

Still, the case of optical illusions inspired me to investigate the philosophy of perception, and its conclusion that we always see things against a background or horizon. Sometimes we can switch back and forth between what we take to be foreground and what background, an experience that has fascinated philosophers interested in the subject: Leibniz wrote about being aboard a moving boat and alternately perceiving the boat or the riverbank as moving, Merleau-Ponty about being on a train alongside another and being able to perceive either one as moving. This helped. When I look at the pendulum, I make it the foreground, and have the surroundings—the stairwell, the building, and so forth—as the background. For this reason I see the plane of the pendulum move, and not the earth. To think that I could see the earth move would be akin to the error of thinking that I could touch the horizon. Are there any circumstances, I wondered, in which I could see the earth move? This would require somehow attaching the plane of the pendulum's swing to a background. It could be done, I figured, and in several ways: What would happen, for instance, if a pendulum were made big enough for me to crawl inside, with windows to look out from? Then my background would be my immediate environment, the bob's interior my own little ark, and I might be able to see everything else as moving around it for the same reason that the ancients saw the heavens revolving around the earth. Or, what would happen if a pendulum were mounted outside on a clear night? Then the background would be the starry heavens, and I might be able to see that the plane of the pendulum's swing was fixed with respect to it and that everything other than these things was in motion.

But this transformed Foucault's pendulum, for me, from a scientific into a philosophical object. It seemed now to reveal the limitations of perception. If what we see is a function of the environment and how we stage what's in the foreground, didn't this mean that something would always be beyond the horizon of perception, unable to be captured? This thought, at once disturbing and liberating, made Foucault's pendulum an instrument of the sublime. Authors such as Burke and Kant have written extensively about the sublime in nature, art, and politics: the sublime involved an experience of the inadequacy of our senses to the presentation of a natural catastrophe. Foucault's pendulum exhibits a different sort of sublime. It exhibits what one might call the scientific sublime, the kind that scientific experiments have insofar as they reveal that nature is infinitely richer than the concepts and procedures with which we approach it. But Foucault's pendulum exhibits yet another kind of sublime as well, having to do with the disclosure of the futility of the expectation of any kind of final answer.

I found this kind of sublime manifested in Umberto Eco's *Foucault's Pendulum*. Casaubon, the narrator, encounters the eponymous device in the Museum of Arts and Trades in Paris, where it causes him to contemplate the deceptive nature of his own perception. If the floor beneath his feet was not still, what was? The experience forces him to embrace a mystery—the mobility of the universe—that hinted at but did not promise the comfort of a fixed point. It exposed him to the temptation to try to control, to manage, the anxiety produced by that mystery by seeking such an (illusory) fixed point. The narrator overhears an earnest and bespectacled boy converse emotionlessly with a female companion about the history and significance of the pendulum. After a moment the couple wanders off, "he, trained on some textbook that had blunted his capacity for wonder, she, inert and insensitive to the thrill of the infinite, both oblivious of the awesomeness of their encounter."[1] Eco's mockery is a little cheap—everyone has sneered at conversations overheard at a museum or concert—but it does illustrate our defenses against the scientific sublime, or, better, our strategies for avoiding the experience altogether.

In Eco's baffling and brilliant book, Casaubon and his comrades encounter a group of fanatics certain of the existence of a grand conspiracy at work in history. They believe that if they could unravel the conspiracy they would have access to knowledge powerful enough to control the world. But the conspiracy does not exist—it is a projection, a desperate and hopeless attempt to find a stable point, an unmoving center, a key to the mysterious: like the pendulum itself. "Even the Pendulum is a false prophet," a companion tells Casaubon. "You look at it, you think it's the only fixed point in the cosmos, but if you detach it from the ceiling of the Conservatoire and hang it in a brothel, it works just the same. . . . It promises the infinite, but where to put the infinite is left to me."[2]

The fanatics who pursue the conspiracy are, therefore, unable to appreciate the sublime, stuck at the last step before it, because they are still convinced of the omnipotent power of understanding. To experience the sublime requires experiencing the futility of this quest—the futility, Kant would say, of trying to bring what they are pursuing into accustomed categories, or measurable limits of comprehension. Casaubon and his knowledgeable associates, though ultimately appreciating the futility of the quest for a fixed point, are never entirely liberated by this realization. They remain riddled with anxiety and vulnerable to the less rarefied and more frightening reactions of the fanatics. Experiencing the sublime, for them, never manages to place the terror at a safe distance.

A few years ago, I returned to the Franklin Institute for the first time since high school thirty-some years ago. The bob is new, and the compass rose has been replaced by a backlit globe. I know much more about the history of this pendulum, its ancestors and siblings. I still don't see the earth turn. Nevertheless, I still find it what I'd call a "deep object," something that guides and disciplines curiosity and fascination into interaction and self-transformation.

Robert P. Crease is Chairman of the Department of Philosophy, State University of New York at Stony Brook.

Perceiving is not a matter of passively allowing an organ—say of sight or hearing—to receive a ready-made impression from without, like a palette receiving a spot of paint. Recognizing and remembering are not matters of stirring up old images of past impressions. It is generally agreed that all our impressions are schematically determined from the start. As perceivers we select from all the stimuli falling on our senses only those which interest us, and our interests are governed by a pattern-making tendency, sometimes called *schema*. In a chaos of shifting impressions, each of us constructs a stable world in which objects have recognizable shapes, are located in depth and have permanence. In perceiving we are building, taking some cues and rejecting others. The most acceptable cues are those which fit most easily into the pattern that is being built up. Ambiguous ones tend to be treated as if they harmonized with the rest of the pattern. Discordant ones tend to be rejected. If they are accepted the structure of assumptions has to be modified. As learning proceeds objects are named. Their names then affect the way they are perceived next time: once labeled they are more speedily slotted into pigeon-holes in the future.

As time goes on and experiences pile up, we make a greater and greater investment in our system of labels.

—Mary Douglas, *Purity and Danger*

SLIME MOLD

Evelyn Fox Keller

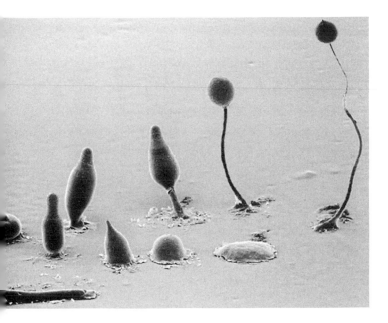

I take an organism as my object, the lowly amoeba-like protist, *Dictyostelium*. In times of plenty, it lives as an individual single-celled organism but, when food supplies are exhausted, it regroups. Then, this one-celled protist becomes part of a complex multicellular motile slug capable of producing fruiting bodies and spores, and of migrating in search of greener pastures where the new spores can germinate.[1] Over the years, it has attracted a great deal of scientific interest, partly because it so elegantly exemplifies a primitive form of biological development, and partly for the paradoxes it embodies. On the one hand, here is a single-celled organism, existing in a population of apparently identical organisms, and on the other hand, it is a part of a differentiated organism assuming a particular role and structure in the larger entity, the multicellular organism. Here is an object that traffics back and forth both between the one and the many and between sameness and difference. For me, this simple being, in its rich ambivalence, has served as an intellectual touchstone, a sustaining object throughout my academic career. Over and over, my work would confront me with a dilemma and this object would resurface to help, offering itself as a model for an entirely new way to think about it. For me, slime mold has unarguably served as an object-to-think-with.

My first encounter with *Dictyostelium* came in 1968. I was working at Cornell Medical College in New York City, looking for ways to fruitfully apply mathematical methods to biological problems. Lee Segel, an applied mathematician from Rensselaer Polytechnic Institute was visiting Cornell for the year, and we teamed up to tackle a couple of problems that looked as though they might be tractable, among these the problem of slime mold aggregation. The onset of aggregation is the first visible step in the process that eventually leads to the cellular differentiation observed in the multicellular organism. Prior to aggregation, there is no apparent difference among cells. But once it occurs, aggregation creates a differential environment for the cells, and therefore it could presumably account for subsequent cell differentiation. The problem is, what sets off the aggregation? Is there some hidden-from-view prior difference, a difference that then serves as the trigger for the development of more elaborate, structured, and clearly visible heterogeneity? Does the onset of organization in fact require the existence of such a preexisting "cause"? Most biologists seemed to think so, and they hypothesized such prior structures under the name of "founder cells." Or was it possible that organization might emerge spontaneously, out of the dynamics of the population as a whole? The first possibility held little appeal for Segel and me: first, it only pushed the question of the origin of heterogeneity further back (where did the founder cells come from?); second, we could find no evidence for the existence of such specialized cells. We set out to demonstrate the feasibility of the second possibility— the notion that organization could emerge from the dynamics of the population as a whole.

The model we developed was a highly simplified— in fact, clearly oversimplified—representation of the actual biological case. Like other mathematical models traditionally employed in the natural sciences, it included just enough of the known biological factors to give rise to the essential phenomenon. At its heart, the model demonstrated that no designated initiator, founder cell, or organizer was required for understanding the advent of aggregation in a uniform distribution of previously undifferentiated cells. We showed that clusters of amoeba would result from the collective dynamics of a population in which a change in external conditions (in this case, depletion of the bacteria that served as their food source) induced a change of state, and indeed, the same change in state, in each individual amoeba. Our account of the onset of differentiation in at least one kind of biological development offered a way to resolve the paradox (how does highly structured difference arise from similarity?) that so sharply divided genetics from embryology. We assumed that it would be of interest to biologists.

But we were wrong—not so much in our model as in our expectations. Biologists, for the most part, showed little interest in our ideas, and despite the absence of evidence, continued to adhere to the belief that founder cells (or pacemakers) were responsible for aggregation. At the time, Segel and I were disappointed and perplexed, but only after ten years had passed did I see how this fact was of interest in itself. What made my new recognition possible was a shift in intellectual mindset and the focus provided by a sharp question posed on the other end of a telephone.

By the early 1980s, I had found a new calling. I had moved my intellectual center of gravity from theoretical physics, molecular biology, and mathematical biology to issues of gender and science. To me, its questions were compelling: How to liberate science from its history of attachment to masculinist ideologies? How to understand the implications of the very different approach to science manifested in work such as that done by the geneticist Barbara McClintock? McClintock had not tried to separate herself from her objects of study—corn cells—to stay more "objective." She imagined herself more "among them," herself reduced to their size, perhaps as a way of becoming one of them.[2]

So, in 1981, slime mold aggregation was far from my mind when, one evening, I received a call from Alan

Garfinkel, a recent convert to mathematical biology. He had recently come across the paper Segel and I had written on the subject, but he had not been able to find any follow-up, either in the form of critique of the work or expansion on the ideas. He asked what was going on. He asked if there was a conspiracy. The call took me aback, not because I thought there had or had not been any conspiracy, but because it immediately brought into focus a problem I had been struggling with around the disparity between McClintock's perceptions and those of her colleagues. I had been wondering why accounts of biological processes that brought such explanatory satisfaction to her colleagues had failed to satisfy McClintock. And vice versa? The sharply worded telephone call reminded me that when I had tried to talk to biologists about our model of slime mold aggregation I had experienced that same wall blocking both interest and understanding.

Suddenly, I saw my own experience as an example of a general phenomenon—a widespread disposition to kinds of explanation that posit a single central causal locus (governor, founder cell, pacemaker)—and that this disposition was crucial in understanding the gap between conventional understandings of biological development as DNA driven and McClintock's own more dynamic proposals. Following David Nanney, I referred to such explanations as "master-molecule theories" and began to wonder why it should be that people tend to find such accounts more natural and conceptually simpler than global, interactive accounts in which causal force is distributed.[3] One possible hypothesis that seemed plausible to me was that we tend to project onto nature our first and earliest social experiences, ones in which we feel passive and acted upon. But in any case, I wrote,

> As scientists, our mission is to understand and explain natural phenomena, but the words *understand* and *explain* have many different meanings.

In our zealous desire for familiar models of explanation, we risk not noticing the discrepancies between our own predispositions and the range of possibilities inherent in natural phenomena. In short, we risk imposing on nature the very stories we like to hear.[4]

Another twenty years have passed since my work on scientists and their preferred narratives of nature. My intellectual preoccupations have shifted again. For more than a decade I wrote about genetics and developmental biology, and today I find myself turning to developmental psychology. Insofar as my focus has remained on the nature of developmental processes per se, the shift has been but a small step, and a new community of intellectual allies was easy to identify. For example, in my work on developmental biology, I had mounted a strong critique of the concept of the "genetic program" (understood as a program for development encoded in the DNA), and I found an entire school of psychology engaging in similar arguments. Its scholars were making strong claims about the value of dynamical systems theory for understanding developmental processes. What drew these psychologists to dynamical systems theory was the language that theory provided for describing the emergence of novel patterns of organization in complex, nonlinear systems, patterns that could not have been predicted from studying the behavior of individual components in isolation.

To me, this had the ring of déjà vu, but much had changed over the years since my early foray into dynamical systems. In particular, familiarity with examples of self-organization—in physics, in computer science, and even in biology (where slime mold aggregation has become a canonical example)—now extends beyond these academic communities. One effect of a more common knowledge of examples of self-organization is that, over the last fifteen years, a series of proposals from different disciplinary quarters urge the reframing of all psychological and biological arguments in terms of dynamical systems.[5]

The word *reframing* here is crucial. Almost all of these proposals suggest alternatives to conventional framings of development in terms of programs (either genetic or developmental), where the very term *program* is seen as implying the unfolding or elaboration of innate capacities. The authors who have taken up the cause of dynamical systems see in this approach an antidote to the prevailing innatism of so much of contemporary writing in biology and cognitive science.

All of these authors seek to redress what they see as the disproportionate emphasis currently placed on internal factors of development; I have enormous sympathy with their concerns. But in this particular intellectual dispute, I found myself jolted out of any comforting alliance by a sharp recollection of slime mold, a call from my past. Slime mold aggregation is illustrative of self-organization in dynamical systems, but does it not equally well illustrate the power of a developmental program embodied in an individual cell? And if it does, then how can it be said that dynamical systems offer an alternative to notions of program? Indeed, I am prompted to ask, can a viable distinction between the two even be made? For me, it cannot. The implications of this heresy are large; they include, for example, the possibility that the battle lines against innatism need, yet again, to be redrawn.

Let us return then to *Dictyostelium*. Our early model was deficient in many ways, yet our central point is still valid: the aggregation of a population of single-celled amoebae (and its subsequent development into multicellular organisms) proceeds spontaneously, without the need for distinctive founder cells; the population emerges as the product of decentralized and local interactions among molecules secreted by individual cells. In other words, despite the elucidation of its genetic organization, *Dictyostelium* has survived as a simple and compelling model of a self-organizing dynamical system. Given the current state of controversy I return to my touchstone to ask: can we therefore say that there is no developmental program for this organism?

That depends, of course, on what we mean by a *program*. The most relevant definition given by the *Oxford English Dictionary* has two parts: first, "a definite plan or scheme of an intended proceedings," and second, "an outline or abstract of something to be done."[6] Both suffer, in this context, from an objectionable degree of anthropomorphism. There is no "intention" guiding the development of an organism, nor is there anywhere an agent "doing" the work. Nevertheless, developmental processes proceed along rather well-defined tracks and conclude with quite predictable outcomes. Remarkably little is left to chance in a developing organism—in fact, it might be said that one of the fundamental characteristics of biological development is the capacity to resist the effects of the myriad vicissitudes the growing embryo inevitably encounters.[7] Thus, to the extent that we can rid the notion of program of its anthropomorphic connotations and think of it simply as a plan or scheme of a proceedings with a definite outcome, a plan that need not be located in a particular structure or homunculus but that may instead be distributed throughout the system, I would argue that the very reliability of most forms of biological development demands the existence of a program.

The key idea of a plan or program for reliable development is that the organism (or machine) must be able to resist the disturbances that can throw it off course, either by suppressing or by adapting to them. In other words, such a program must have contingency built into it—instructions, if you will—for how to respond to a range of different kinds of input. In the case of slime mold, the single cell needs to have a change-of-state plan—a plan for changing certain key cell parameters when the food runs out. More sophisticated organisms are equipped with programs (or built-in strategies) for changing state in response to changes in a far larger set of parameters. A computer program may not be such a bad image after all, but think of it as a program for survival. Such a program (or strategy) no more requires anticipation than does any other function that has been evolved by the process of natural selection.

The main challenge for the notion of a program located inside the individual cell lies elsewhere: if we are to grant the existence of a program inside the individual (undifferentiated) slime mold cell, that program must not only allow for the change of state in that cell to be triggered by starvation but also for the reproduction of the cell. Without reproduction, there will be no population, and without a population of cells, there can be no aggregation. But after fifty years of work, most of which has been inspired by John von Neumann's early efforts, this problem too seems to have been resolved for programs. Today, programs for reproduction—in virtual even if not in physical space—have become ubiquitous. There are of course still problems, and these problems—primarily having to do with the programs' lack of robustness—are largely responsible for their confinement in virtual space. In this sense, real organisms remain far ahead.

As an object-to-think with, slime mold has proven to be an immense resource. I am grateful for this opportunity to pay it homage. Some have resisted its message, but there is, too, the danger of overusing it. There are limits to what it has to teach us. The particular route to multicellularity it manifests is, after all, a rather primitive one; furthermore, it bears little resemblance to the developmental process by which most complex organisms come into being. Slime mold may be equipped with a program for adapting to the scarcity of food, but the developmental programs of higher organisms must deal with a far larger range of variability, and evolution has equipped them with an extraordinary repertoire of ways of adapting to such variability. The world challenges them anew each and every day and in ways that could not possibly be met with a single tool, or even a few, or perhaps not even with a finite number of tools. Slime mold, in its capacity for self-organization, illustrates one strategy for survival, and it is undoubtedly a versatile and fertile object-to-think with. But ultimately more complex living beings find the need of a far larger repertoire of strategies than this little organism can possibly be expected to display.

Evelyn Fox Keller is Professor Emeritus of History and Philosophy of Science in the Program in Science, Technology, and Society at MIT.

WHAT MAKES AN OBJECT EVOCATIVE?

Sherry Turkle

What makes an object evocative?[1] As I write, *Bodies,* an exhibition of preserved humans from China, is on tour internationally. Its objects, poised between death and new animation, raise questions about the sanctity of what has lived, the nature of art, and the human beings who once were the objects on display. Thinking about the uncanny, about thresholds and boundaries helps us understand these objects with their universal powers of evocation.

And yet, the meaning of even such objects shifts with time, place, and differences among individuals.[2] Some find the preserved bodies the fearsome creatures of night terrors. For others, they seem almost reassuring, an opportunity to contemplate that although death leaves matter inert, a soul may be eternal.

To the question "What makes an object evocative?" this collection offers pointers to theory (presented as epigraphs) and the testimony of its object narratives, voices that speak in most cases about familiar objects—an apple, an instant camera, a rolling pin. One role of theory here is to defamiliarize them. Theory enables us, for example, to explore how everyday objects become part of our inner life: how we use them to extend the reach of our sympathies by bringing the world within.

As theory defamiliarizes objects, objects familiarize theory. The abstract becomes concrete, closer to lived experience. In this essay I highlight the theoretical themes of each of the six parts of this collection (with special emphasis on objects and the inner life) in the hope that theory itself will become an evocative object. That is, I encourage readers to create their own associations, to combine and recombine objects and theories—most generally, to use objects to bring philosophy down to earth.

It was made of two wheels and an axle, with a pin hanging down from the middle of the axle (not quite hitting the ground), and a string at the end of the pin.
—Mitchel Resnick, "Stars"
Objects of Design and Play

Objects help us make our minds, reaching out to us to form active partnerships. Mitchel Resnick's pull-toy, a wooden car on a string, embodied a paradox: "Since the string is attached to the end of the pin, it seems that the pin should come toward you. At the same time, it seems that the wheels should come toward you. Both can't be true." Resnick had been shown the pull-toy in his high school physics class; he brought the idea of the toy car home with him, but more than this, he brought home the notion of paradox itself. He took apart his own, familiar toys for parts that enabled him to rebuild the pull-toy in his fashion, and even when he had come to understand its mysteries, he continued tinkering: "Even after I 'knew' the answer, I loved tugging on the string and thinking about the paradox." The object took on a life of its own. "No ideas but in things," said the poet William Carlos Williams.[3] And the thing carries the idea.

The anthropologist Claude Lévi-Strauss would say that as Resnick made and remade the pull-toy he was becoming a scientist, more specifically, a *bricoleur,* a practitioner of the science of the concrete. Bricolage is a style of working in which one manipulates a closed set of materials to develop new thoughts.[4] Lévi-Strauss characterizes the primitive scientist as a bricoleur, but modern engineers, too, use this style.[5]

From our earliest years, says the psychologist Jean Piaget, objects help us think about such things as number, space, time, causality, and life.[6] Piaget reminds us that our learning is situated, concrete, and personal. We invent and reinvent it for ourselves. As Resnick plays with pull-toys, he is learning to see himself as capable of inventing an idea, and he is changing in other ways as well. He is learning to be more at home with uncertainty and with his own object attachments.

Object play—for adults as well as children—engages the heart as well as the mind; it is a source of inner vitality. Resnick reminds us of how his mentor, the mathematician and educator Seymour Papert, considered the lessons of his childhood object: gears. An intimate connection with gears brought Papert in touch with ideas from mathematics. As Papert put it: "I fell in love with the gears."[7] Far from being silent companions, objects infuse learning with libido.

Another of Papert's students, Carol Strohecker, proposes knot-tying as a microworld that similarly combines ideas and emotions. Here, I pair her essay with the writing of Lévi-Strauss, a connection that puts the focus on the cognitive. But reading Strohecker's narrative from a psychoanalytic perspective shifts the emphasis to emotion and the particular needs of individuals.[8] In *Playing and Reality*, Winnicott describes how one of his patients, a seven-year-old boy, becomes obsessed with string in response to the anxiety of being separated from his hospitalized mother. At each hospitalization, the boy turns to string play as solace, as a way of coping with her absence.[9]

Similarly in Strohecker's "Knot Lab," ten-year-old Jill, a child of a difficult divorce, is preoccupied with tying down the ends of string as she works, using tape, nails, and tacks to keep her knots in place. For Jill, knots are a way to think through her personal situation. Herself at loose ends, Jill is comforted by securing knots in transition. When she builds a knot exhibit that enables passersby to play with the back-and-forth movement of a True Lovers' Knot, her label for the knot concludes with the phrase "please pull me." Strohecker hears Jill speaking through the knots: "Notice how I am suspended by two knots, one that anchors me and one that holds me. Notice how I am two knots, waiting to be pulled this way and that. I understand being pulled; it is something that I know. Allowing others to pull me is a purpose that I serve."

My datebook and its events had their own esoteric language. Familiar venues, organizations, and individuals were noted in tiny writing and abbreviations that only I could decipher.
—Michelle Hlubinka, "The Datebook"
Objects of Discipline and Desire

Michelle Hlubinka writes about her datebook and her first timepiece—a Mickey Mouse watch that she received on a family vacation when she was four: "Having the watch, I entered a society not just of time-keepers, but time-managers. And I became good at it, perhaps too good at it."

You think you have an organizer, but in time your organizer has you. The organizer is one of many day-to-day technologies that concretize our modern notion of time. The historian of technology Lewis Mumford examines how the invention of the clock by monks in the Middle Ages transformed social life and subjectivity.[10] Clocks produced time as discrete units, making possible a new way of thinking. Before clocks, there was day and night, morning, mid-day, and evening. Soldiers showed up for battle at dawn. After clocks, there were minutes and seconds. Industrialization needed a clock-produced world of measurable sequences and synchronized action. Capitalism depends on regimenting human time and human bodies.

Our clocks and datebooks do more than keep us on time. Objects function to bring society within the self.

The historian Michel Foucault provides a framework for thinking about how objects such as Hlubinka's watch and datebook serve as foundations of "disciplinary society."[11] In modern times, social control does not require overt repression. Rather, state power can be "object-ified."[12] Every time we fill out a medical questionnaire or take a pill, we are subjects of social discipline. And every time we enter appointments in our datebook, we become the kind of subjects that disciplinary society needs us to be.[13]

When literary theorist Roland Barthes writes that the objects of disciplinary society come to seem natural, what is most important is that what seems natural comes to seem right. We forget that objects have a history. They shape us in particular ways. We forget why or how they came to be. Yet "naturalized" objects are historically specific. Contemporary regimes of power have become capillary, in the sense that power is embodied in widely distributed institutions and objects.

From this perspective, Gail Wight's object—the antidepressant medication she calls "Blue Cheer"—produces a patient, just as Hlubinka's datebook produces a time-keeper and time-manager. At the start of Wight's narrative about her pills, she has a sense of herself as an unhappy artist. Soon, psychiatry recasts her identity: she is a broken biological mechanism, but one that medicine can fix. Over time, Wight does not need the presence of a physician to reinforce her medical identity. Over time, the pills alone can do the job.[14]

Eden Medina, like Wight, has her body disciplined. In Medina's case, the social demands are embodied in her shoes. The ballet slippers that haunt Medina communicate the shape of the body to which they want to belong: the ideal dancer's body, conforming to the socially constructed conventions of ballet. Toe shoes put Medina in touch with body practices that teach how the flesh disappoints and how it needs to be disciplined and denied.[15]

Although it looked like a Braun transistor radio, this object never produced sound. I asked the boy about it and

he said: "It can't play music, but I sing when I carry it.
One day I'll have a real one."
—Julian Beinart, "The Radio"
Objects of History and Exchange

Julian Beinart saw a new object, a mute radio made of wood, and then he could not stop seeing it. His hometown of Durban, South Africa, revealed itself to be rich in technological objects fashioned from the raw materials of an impoverished culture. There were bicycles made from beer cans, cars from bent wire, radios from wood— all technologies of everyday life copied as pure form.

As Beinart found these objects, he saw people and social relationships of which he had been previously unaware. The mute radio and its cousins changed the people who made them and Beinart who discovered them. The mute radio, with no instrumental purpose, was free to serve as commentary on possession and lack, on power and impoverishment.

In a famous passage on commodities, Karl Marx describes how when wood is transformed into a table, it remains an ordinary, sensuous thing. But when the table becomes a commodity in a market system, the object comes alive: it "stands on its head and evolves out of its wooden brain grotesque ideas far more wonderful than if it were to begin dancing of its own free will."[16] Like Marx's commodities, Beinart's wooden radio comes alive as it embodies relationships to power. Yet the wooden radio subverts itself as a commodity and reveals the social relations that commodities are designed to hide.

The social theorist Marcel Mauss, too, describes the animation of objects: gifts retain something of their givers.[17] As people exchange objects, they assert and confirm their roles in a social system, with all its historical inequalities and contradictions. A gift carries an economic and relational web; the object is animated by the network within it.

From the perspective of the philosopher Jean Baudrillard, the mute radio reveals something profound about the social role of all the radios that can speak. He describes how commodities cultivate desires that support the production and consumption capitalism requires.[18] This process keeps the dominant ideology alive. It becomes invisible and alienates from the real. In such a system, normal radios are taken for granted. But when radios are remade in wood or throw-away tin, the invisible is made visible. In wood, a radio is subversive, a potent actor.

David Mitten finds a Native American axe head that also speaks to him in a subversive way. It subverts his sense of distance between himself and those who came before him, a theme of the writings of Bruno Latour, with whom his essay is paired. For Latour, objects speak in a way that destroys any simple stories we might tell about our relations to nature, history, and the inanimate; they destroy any simple sense we might have about progress and our passage through time.[19] Mitten says that when he picked up the axe head, the landscape of his ancestry exploded around him, demanding that it be placed in history, in nature, and in the social lives of the people who had and used it. More than this, Mitten knows that he will part with the axe head only in death, when his daughter will inscribe his life into stories about it.

A bunny with a soft cotton collar less than half-an-inch wide was named Collar Bunny. . . . He had a small plastic rattle inside his body, and when he sat, the stuffing in his arms made them stick out to the sides.
—Tracy Gleason, "Murray: The Stuffed Bunny"
Objects of Transition and Passage

D. W. Winnicott called "transitional" the objects of childhood that the child experiences as both part of the self and of external reality. Collar Bunny (later renamed "Murray") is such an object.

He belongs to Tracy Gleason's younger sister, Shayna. Whatever Shayna imagines herself doing or thinking ("like dressing herself and hopping on one foot and telling a silly joke") can first be "tried on" as bunny thoughts and actions.

Winnicott writes that the transitional object mediates between the child's sense of connection to the body of the mother and a growing recognition that he or she is a separate being. When Shayna starts preschool and its rules insist that Murray cannot accompany her, she is challenged to invent ways of bringing him along. Her solution is to invest Murray with new powers. He develops the ability to read Shayna's mind and intuit her every emotion. In doing so, Murray makes it possible for separation to be not-quite separation. Transitional objects let us take things in stages.

The transitional objects of the nursery—the stuffed animal, the bit of silk from the baby blanket, the favorite pillow—all of these are destined to be abandoned. Yet they leave traces that will mark the rest of life. Specifically, they influence how easily an individual develops a capacity for joy, aesthetic experience, and creative playfulness. Transitional objects, with their joint allegiance to self and other, demonstrate to the child that objects in the external world can be loved. Winnicott believes that during all stages of life we continue to search for objects we can experience as both within and outside of the self.

It is in these terms, as an object in the space between self and surround, that Judith Donath speaks of her much-beloved 1964 Ford Falcon. She inhabits the car like a "skin"; it connects her to her mother, its first owner, and to her children, for whose safety she abandons it. It brings her the joy of an object that traffics, in her words, "between the outside world and the inner self."

Donath's essay is paired with the writing of the anthropologist Igor Kopytoff, who explores objects in
terms of their life spans, a perspective that encourages
us to look at the biography of an object alongside that of
a person. Through Donath's sensitivity to the Falcon's
cultural biography, she was better able to understand
her own. When Donath rides the Falcon as a child in the
1970s, it is a bourgeois suburban object. When it re-
appears in New York's East Village in the 1980s, the Falcon has been transformed into the neighborhood "cool
car." By the 1990s in Cambridge, Massachusetts, the
car is exotic and glamorous, congruent with Donath's
desire to stand out as a graduate student. "No matter
how dully mundane I felt, in the Falcon I was the Driver
of that Cool Car."

Winnicott situated his transitional objects in play,
which he saw as an intermediate space, a privileged zone
in which outer and inner realities can meet.[20] For William J. Mitchell, born in the outback of Australia, the
train to Melbourne provided such a space.

The train is the backdrop for a rite of passage, a
time of transition that the anthropologist Victor Turner
has characterized (for individuals and cultures) as "liminal" or threshold time.[21] For Turner, these times of transition are characterized by the crystallization of new
thought and the production of new symbols.

On the Melbourne train, Mitchell is taken from one
physical space (his small village in the Australian bush)
to another (the cosmopolitan Melbourne), and he is also
taken toward a new identity. He writes: "Each warmly lit
carriage interior was a synecdoche of urbanity—an encapsulated, displaced fragment of the mysterious life
that was lived at the end of the line." Within the liminal
space, the self is porous. In train space, Mitchell is open
to new associations, sights, and sounds: "And there
were wondrous cabinets of curiosities, with friezes of
large, sepia photographs over the seats."

In liminal space, Mitchell brings books, words, and objects within his expanding sense of self. It is on the train that he first realized that he can read.

> "It was on a train, long before I was reluctantly dragged off to school, that I first realized I could read . . . words in memorable sequence, the beginnings of narrative. . . . As the years went by, and I made myself into an architect and urbanist, I began to understand that objects, narratives, memories, and space are woven into a complex, expanding web—each fragment of which gives meaning to all the others."

Mitchell's essay, rich in its discussion of language, is paired with an excerpt from the literary theorist Roland Barthes, whose reflections on objects, language, and identity (he writes of "language lined with flesh") also resonate with those of David Mann, writing about the transitions facilitated by the *World Book Encyclopedia* he received as a child.[22]

Far more than a vehicle for the transfer of information, Mann describes the encyclopedia as a means of access to language:

> Its pictures came to life in my mind, parsed into nouns and danced through grammar to the music of verbs. By the time I was four it had taught me to read. Not through my family but through these volumes language became a part of me, the book of the world opened to me and I myself opened to the world as I might otherwise never have done.

Mann and Mitchell make language itself a liminal object, standing outside and within the self, a vehicle for bringing what is outside within.

Mann's description of a self constituted by language is paired with a text by the psychoanalyst Jacques

Lacan. Lacan believes that to talk of "social influences" on the individual neutralizes ole of Freud's most important contributions: the recognition that society doesn't "influence" autonomous individuals, but comes to dwell within them with the acquisition of language.[23]

Lacan's theory allows for no real boundary between self and society. People become social with the appropriation of language. You and language become as one. There is no natural man. Lacan's narrative of how language comes to "inhabit" people during the Oedipal phase opens out to larger questions about how we build our psyche by bringing things within. Nowhere is this more in evidence than when we consider what we bring within at a time of loss.

The logo boasts "Globe Trotter," echoing my grandmother's love of travel. With her newfound liberty after her husband and children had gone, she began to discover the world. . . . But this suitcase is new; she had been saving it for one final trip.
—Olivia Dasté, "The Suitcase"
Objects of Mourning and Memory

After her grandmother's death Olivia Dasté packs the old woman's suitcase one last time. A sweater, a handkerchief, a teacup are lovingly arranged in the suitcase. Dasté is afraid to open the suitcase too soon: "[I]t feels dangerous to open it. Memories evolve with you, through you. Objects don't have this fluidity; I fear that the contents of the suitcase might betray my grandmother." But after two years, mourning has done its work. Dasté holds a fragrant red sweater to her face and knows she doesn't have to. Dasté has internalized her grandmother's spirit. "I smile. I am with her in Bordeaux and we have all the time in the world."

In *The Year of Magical Thinking,* Joan Didion describes how material objects may look during the mourning process.[24] After her husband's death, Didion cannot bring herself to throw away his shoes because she is convinced that he may need them. This is the magical thinking that is associated both with religious devotion and the "illness" of mourning. With time, Freud believed, the true object, the lost husband, comes to have a full internal representation.[25] This completes the formal process of mourning; it is only at this point that the shoes can be relinquished. They have served a transitional role.

Susan Pollak, too, begins her narrative of loss with an echo of the tactile—brought back by the way a rolling pin evokes her grandmother's kitchen, the safe place of Pollak's childhood.[26] Pollak's thoughts then go to baking and to the evocative object of Marcel Proust, perhaps the most famous evocative object in all literature. Proust's object is the small cookie called a madeleine. When dipped in tea, the taste of the madeleine brings Proust's character back to his youth, to a country home in Combray, and to his aunt Albertine. Finally, the madeleine opens him to "the vast structure of recollection."[27]

"Never underestimate the power of an evocative object," says Pollak. As a practicing psychotherapist, she is interested in objects for more than evocation. She argues, following Winnicott, that transitional objects can heal. Pollak tells the story of a patient, Mr. B., who was not able to mourn his father until he found the "half-moon" cookies his father had bought for the family when Mr. B. was a child. At that time, money had been tight and his father had only been able to buy day-old cookies. When Pollak's patient went back to his old neighborhood and found the bakery from his childhood, he bought a dozen fresh half-moon cookies. They were unfamiliar, almost displeasing. He had to wait until they were a day old in order to savor them. Only the taste and texture of his childhood could reestablish his lost connection. After finding the cookies he was able to talk to his children about their grandfather. He was able to recall his father's acts of generosity and to think sympathetically about why his father had needed alcohol to endure. The cookie facilitated mourning. Mr. B., a novelist, long blocked in his writing, was able to begin a new novel. For him, as for Proust, memory passed through the body.[28]

Pollak reminds us that Proust himself makes a connection that Winnicott would wholeheartedly endorse. Toward the end of *Remembrance of Things Past*, he says: "Ideas come to us as the successors to griefs, and griefs, at the moment when they change into ideas, lose some part of their power to injure our heart."[29]

My rocks are un-rock-like. They are plain limestone contradicting itself. The most earthy and banal material transcends itself to become exotic.
—Nancy Rosenblum, "Scholars' Rocks"
Objects of Meditation and New Vision

In a narrative in which ideas are successors to grief, Nancy Rosenblum, the widow of a sculptor who collected Chinese scholars' rocks, asks, "How can a rock be a man?"

Scholars' rocks are found in nature, then mounted on meticulously worked bases. The bases transform the rocks into things that are made as well as found, objects that invite reflection on the boundary between nature and culture. Says Rosenblum: "They have the power to provide an effortless, aesthetic experience of mystery. Of infinity in a finite space. Of transformation. Just by looking. Without philosophy."[30]

The rocks displace scale, time, and authorial intent. They are classically liminal objects in Turner's sense: betwixt-and-between categories, the rocks challenge the categories themselves. As Rosenblum puts it, "Gaze at a stone and it disorients."

In traditional rites of passage, participants are separated from all that is familiar. We saw that this makes them vulnerable, open to the objects and experiences of their time of transition. The contemplation of liminal objects can make us similarly vulnerable. In their disorienting qualities, in the way they remind us of the mundane yet take us away from it, scholars' rocks share something of what Freud called the uncanny, those things "known of old" yet strangely unfamiliar.[31]

In his writing on the uncanny, Freud analyzes the etymology of the German words *heimlich* and *unheimlich,* roughly the homelike and familiar and the eerie and strange. The two words seem to be the opposite of each other, suggesting that the eerie is that which is most unfamiliar. But among the meanings of *heimlich* (familiar) is a definition close to its opposite: it can mean concealed or kept out of sight. Heimlich has a "double." By extension, Freud argues, our most eerie experiences come not from the exotic, but from what is close to home. Uncanny objects take emotional disorientation and turn it into philosophical grist for the mill.

In this collection, Jeffrey Mifflin, the curator at Boston's Massachusetts General Hospital, uses a 2,600 year-old mummy to ponder ultimate questions: "He had been flesh and blood and bone, and the flesh and bone were still there. His senses had once worked as mine now did. His mind was gone, but neither would I live forever."

Mifflin's mummy frightens him even as it grows in his affections. The man who became the mummy was Padihershef, a stonecutter who lived near Thebes during the Saite Period (XXVI Dynasty) and died in his late forties. His specialty was cutting stone to make tombs. Mifflin begins to identify with Padihershef. When Mifflin opens the mummy's exhibit case and smells the embalming spice and chemicals, he is not overtaken by their pungency, but by the thought that Padihershef's own friends would have smelled something quite similar as they closed his coffin.

Mifflin calculates the generations between himself and the mummy, in his estimate about 130, and he wonders if his "distant progenitors in Britain were mining tin or slicing blocks of peat at the same time that Padihershef was chiseling out tombs in Egypt?" Mifflin thinks about his own uncertainties about religion and the afterlife in relation to Padihershef's probable certainties. Mifflin measures their lives against each other, each seeking to find a place in history and in his generation.

As a curator, Mifflin compares the untidy, chaotic spaces in museum back rooms and the meticulous presentations in the front rooms where all is tidy and ordered. The contrast reveals something too often hidden: we tend to present "front room" knowledge as "true." But its certainties are constructed. We make up a clean story to mask our anxieties about the chaotic state of the little that we know. Chaos compels its opposite: "the orderly presentation of supposed facts" to which Mifflin feels disconnected. He fears that he will always be blocked in his ability to experience certainties by his access to their opposite—his experience in the dirty back rooms. Yet it is the contrast between the front and back rooms that leads Mifflin to a new appreciation of the complexity of knowledge.

In *Purity and Danger,* the anthropologist Mary Douglas examines the evocative power of such contrasts, focusing on how the tension between order and disorder is expressed through our relationship to dirt and pollution.[32] Order is defined in terms of dirt, or that which is not polluting. And dirt is defined in terms of order. Societies create the classification "dirt" to designate objects that don't fit neatly into their ways of ordering of the world.

This collection ends with Evelyn Fox Keller's reflections on her life in science, a narrative about the power of order-disrupting ("dirty") objects to provoke meditation and new vision. Keller takes slime mold as her object, an object full of paradoxes: "In times of plenty, it lives as an individual single-celled organism but, when food supplies are exhausted, it regroups. . . . [It] traffics back and forth both between the one and the many and between sameness and difference."

Turner and Douglas help us see things on the boundary, such as slime mold, as both disruptive and as sources of new ideas. Indeed, for Keller, the "betwixt-and-between" slime mold not only becomes an object-to-think-with for thinking about processes within cells, it becomes a way to think about the politics of science.

In the late 1960s, most biologists argued that slime mold goes from being a unicellular to a multicellular organism, following a signal given by "founder cells." In a 1968 paper, Keller and biologist Lee Segel disagreed. They suggested that changes in the slime mold's state followed from the dynamics of the cell population as a whole. There was no command and control center that took charge of the process. Biologists resisted this suggestion. Keller says: "[D]espite the absence of evidence, [biologists] continued to adhere to the belief that founder cells (or pacemakers) were responsible for aggregation."

Two decades later, while working on a biography of the geneticist Barbara McClintock, Keller again faced the resistance of biologists—this time to a style of doing science. Canonical scientific methods insisted on the researcher's distance from the object of study, but McClintock wanted to be close to her objects, among the corn cells of her research. She imagined herself like a modern-day Alice, brought to their scale in order to feel more a part. Her colleagues in biology were not impressed. Keller began to identify with McClintock. Like her subject, when Keller had looked at cells, she had seen social and decentralized processes. Keller comes to see her career and McClintock's as illustrative of how biology rejects theories that challenge the dogma of single and centralized causal factors.

As Keller wonders why people find causal accounts so compelling, she considers explanations that draw on the Freudian tradition. There, our earliest, profoundly bonded, connections to the world are interrupted by a sudden experience of separation. Keller hypothesizes that "we tend to project onto nature our first and earliest social experiences, ones in which we feel passive and acted upon." Whether or not this particular hypothesis is true, she says, a more general point certainly is: scientists were not open to the "discrepancies between our own predispositions and the range of possibilities inherent in natural phenomena. In short, we risk imposing on nature the very stories we like to hear."[33]

What are the stories we like to hear? Keller suggests that they are often the ones that confirm us in comfortable ways of thinking. But theory can help us to see things anew.

Until now, I have discussed physical objects that engender intimacy. What becomes of this intimacy when people work with digital objects?

Any response needs to be complex, as is apparent in the contrast between two essays in this collection. Mitchel Resnick describes his StarLogo program that brings its users to an encounter with ideas about emergent phenomena, much as the concrete objects of Piaget's day put children in touch with ideas about counting and simple categorization. His goal is to have the computer enable a new kind of learning. Yet Susan Yee's testimony about work in a digital archive suggests aspects to life on the screen that may be inherently alienating.

Yee, an architect, begins her relationship with Le Corbusier through the physicality of his drawings. As she works in the Le Corbusier archives in Paris, his original blueprints, sketches, notes, and plans are brought to her in long metal boxes. Le Corbusier's handwritten notes in the margins of his sketches, the traces of his fingerprints, the smudges, the dirt, all of these encourage Yee's identification with the designer. To Yee, the most "miraculous" moment in the physical archive is finding the little colored paper squares that Le Corbusier used to think through his design for the Palace of the Soviets. Yee says that she could imagine Le Corbusier "fiddling" with the design elements, moving them around, considering different shapes and volumes as he worked. The little bits of colored paper connect Yee to his process. Delighted, Yee "fiddles" with them too. The bricolage of the master is re-experienced in the bricolage of the student. As it happened, Yee was visiting the Le Corbusier archive at a dramatic moment, the day it was converted from physical to virtual space. The philosopher Jacques Derrida sees such transitions as "transforming the entire public and private space of humanity."[34] For one thing, while any archive is a selection of material that erases what has been excluded—the digitized archive goes a step further. Its virtuality insures another level of abstraction between its users and what has been selected. It brings to mind Derrida's writing about the word processor where "erasure" is central to his concerns: "Previously, erasures and added words left a sort of scar on the paper or a visible image in the memory. There was a temporal resistance, a thickness in the duration of the erasure. But now everything negative is drowned, deleted; it evaporates immediately, sometimes from one instant to the next."[35]

Derrida's meditation on erasure brings us back to what troubled Yee in the archive. She is aware that, digitized, the Le Corbusier archives will be available to scholars all over the world and be protected from wear and tear. Yet, when the archive is digitized, Yee experiences the loss of her connection to Le Corbusier: "It made the drawings feel anonymous," she says. More important, the digitized archives make Yee feel anonymous. She is grateful for her own position in a generation of architects that knows drawing by hand as well as by computer; her narrative captures an anxiety that digital objects will take us away from the body and its ways of understanding.

Through Yee's essay on the archive, this collection engages the problem of virtuality and its discontents. Yet her cautionary essay must be read in relation to other narratives about computational objects—represented by the promise and enthusiasm of Resnick's writing, as well as that of Howard Gardner, Trevor Pinch, and Annalee Newitz—that suggest how digital objects engage us in new and compelling ways.

Indeed, in Newitz's description of her laptop computer, the flickering screen does not appear cold and abstract, but is integrated into her sense of herself. Her experience of the laptop is reminiscent of how Joseph Cevetello, a diabetic, talks about his glucometer, a device for measuring blood sugar. Cevetello notes how over time his glucometer becomes more than companion: the glucometer "has become me." Moment to moment, its output determines his actions. He lances his finger, readies an insulin injection, and waits "for my meter to tell me what to do." The laptop, like the glucometer, is experienced as co-extensive with the self. Newitz feels so close to her laptop that she cannot tell where it leaves off and she begins. Her self-understanding depends on analyzing the flows and rhythms that pass between herself and the machine. In bed, Newitz remembers not to let the blankets cover the computer's vents so it does not overheat. She is at one with her virtual persona: "I was just a command line full of glowing green letters."

Cevetello and Newitz have achieved couplings so intimate between themselves and their objects that we might characterize them as cyborg.[36] In the cyborg world we move beyond objects as tools or prosthetics. We are one with our artifacts. And in the cyborg world, the natural and the artificial no longer find themselves in opposition. Says the historian of science Donna Haraway:

"Any objects or persons can be reasonably thought of in terms of disassembly and reassembly."[37] No object, space, or body is sacred in itself: "Any component can be interfaced with any other if the proper standard, the proper code, can be constructed for processing signals in a common language."[38] Newitz still has to carry her laptop around, but the day is not far off when computation will become part of our bodies, beginning with chips to improve our sight and hearing. Cevetello anticipates the day when his glucometer will be available as an implant; it will provide a digital readout directly sensed by his body.

Once we see life through the cyborg prism, becoming one with a machine is reduced to a technical problem of finding the right operating system to make it (that is, *us*) run smoothly. When we live with implanted chips, we will be on a different footing in our relationships with computers. When we share other people's tissue and genetic material, we will be on a different footing with the bodies of others. Our theories tell us stories about the objects of our lives. As we begin to live with objects that challenge the boundaries between the born and created and between humans and everything else, we will need to tell ourselves different stories.

Notes

Sherry Turkle | *The Things That Matter*

1. Claude Lévi-Strauss, *The Savage Mind*, trans. John Weightman and Doreen Weightman (Chicago: University of Chicago Press, 1966).
2. See Sherry Turkle, *Life on the Screen: Identity in the Age of the Internet* (New York: Simon & Schuster, 1995), 54–56.
3. See, for example, Jean Piaget and Barbel Inhelder, *The Growth of Logical Thinking from Childhood to Adolescence*, trans. Anne Parsons and Stanley Milgram (New York: Basic Books, 1958).
4. Lévi-Strauss, *The Savage Mind*, 16ff.
5. In science studies, two groundbreaking ethnographies that showed the power of the concrete were Bruno Latour and Steven Woolgar, *Laboratory Life: The Social Construction of Scientific Facts* (Princeton, N.J.: Princeton University Press, 1986 [1979]), and Karin Knorr Cetina, *The Manufacture of Knowledge: An Essay on the Constructivist and Contextual Nature of Science* (Oxford, Pergamon Press 1981). As for Nobel laureates, Richard Feynman wrote extensively about his everyday tinkering in *Surely You're Joking, Mr. Feynman* (New York: W. W. Norton, 1981).
6. One example of feminist scholarship that focuses on a scientist's profound connection with her objects is Evelyn Fox Keller's biography of Barbara McClintock, *A Feeling for the Organism: The Life and Work of Barbara McClintock* (San Francisco: W. H. Freeman, 1983). Other early feminist contributions include Ruth Bleier, ed., *Feminist Approaches to Science* (New York: Pergamon, 1986); Carol Gilligan, *In a Different Voice: Psychological Theory and Women's Development* (Cambridge, Mass.: Harvard University Press, 1982); Sandra Harding and Merrill B. Hin-

tikka, eds., *Discovering Reality: Feminist Perspectives on Epistemology, Metaphysics, Methodology, and Philosophy of Science* (London: Reidel, 1983).

7. To take only a few examples, see Arjun Appadurai, ed., *The Social Life of Things: Commodities in Cultural Perspective* (Cambridge: Cambridge University Press, 1988); Bill Brown, ed., *Things* (Chicago: University of Chicago Press, 2004); Lorraine Daston, ed., *Things that Talk: Object Lessons from Art and Science* (New York: Zone Books, 2004); Mihaly Csikszentmihalyi and Eugene Rochberg-Halton, *The Meaning of Things: Domestic Symbols and the Self* (Cambridge: Cambridge University Press, 1981); Karin Knorr Cetina, *Epistemic Cultures: How the Sciences Make Knowledge* (Cambridge, Mass.: Harvard University Press, 1999); Bruno Latour, *We Have Never Been Modern*, trans. Catherine Porter (Cambridge, Mass.: Harvard University Press, 1993); and Edward Tenner, *When Things Bite Back* (New York: Knopf, 1996). For more resources, see the bibliography following these essays.

8. For Freud's description of loss leading to internalization and mental representation, see Sigmund Freud, "Mourning and Melancholia," in *The Standard Edition of the Complete Psychological Works of Sigmund Freud*, ed. and trans. James Strachey et al. (London: Hogarth, 1953–1974), vol. XIV, 239–258. As it is for people and the wishes associated with them, so I believe it is for objects: they become powerful in our psychic lives as we bring them within us, along with their associations, emotional and intellectual.

9. Ibid., "The Uncanny," vol. XVII , 219–252. In their provocation to discourse, uncanny objects contrast to fetish objects that are stand-ins for thoughts that cannot be expressed; they take the place of what cannot be spoken; "Fetishism," vol. XXI, 152–157.

10. D. W. Winnicott referred to such objects as transitional. See *Playing and Reality* (New York: Routledge, 1989 [1971]).

11. See Victor Turner, *The Ritual Process: Structure and Anti-Structure* (Chicago: Aldine, 1969).

Carol Strohecker | Knots

1. See Carol Strohecker, "Elucidating Styles of Thinking though Learning about Knots," in *Constructionism: Research Reports and Essays, 1985–1990, by the MIT Epistemology and Learning Group, the MIT Media Laboratory,* ed. Idit Harel and Seymour Papert (Norwood, N.J.: Ablex, 1991), 215–233, and Strohecker, "Why Knot," Ph.D. diss., MIT, 1991, and "Understanding Topological Relationships through Comparisons of Similar Knots," *AI & Society: Learning with Artifacts* 10 (1996): 58–69.

Susan Yee | The Archive

1. See for example, Sherry Turkle, *The Second Self: Computers and the Human Spirit* (Cambridge, Mass.: MIT Press, 2005 [1984]), 18–19.

Mitchel Resnick | Stars

1. Portions of this essay previously appeared in my book *Turtles, Termites, and Traffic Jams: Explorations in Massively Parallel Microworlds* (Cambridge, Mass.: MIT Press, 1994).

2. Seymour Papert, "The Gears of My Childhood," in *Mindstorms: Children, Computers, and Powerful Ideas* (New York: Basic Books, 1980), vi–viii, and Sherry Turkle, *The Second Self: Computers and the Human Spirit* (Cambridge, Mass.: MIT Press, 2005 [1984]).

3. See Friedrich Froebel, *The Education of Man,* trans. and annot. W. N. Hailman (Mineola, N.Y.: Dover Publications, 2005 [1892]).

Joseph Cevetello | The Elite Glucometer

1. Manfred Clynes and Nathan Kline, "Cyborgs and Space," *Astronautics* 14 (1960): 26–27 and 75–76.

2. Rainer Maria Rilke, *Letters to a Young Poet,* trans. Stephen Mitchell (New York: Vintage, 1986 [1906]), 84.

Michelle Hlubinka | *The Datebook*

1. Benjamin Franklin, *The Autobiography of Benjamin Franklin,* Mineola, N.Y.: Dover, 1996, 63.

Annalee Newitz | *My Laptop*

1. John J. Ratey, *A User's Guide to the Brain: Perception, Attention, and the Four Theaters of the Brain* (New York: Vintage, 2002).

Julian Beinart | *The Radio*

1. Jean Baudrillard, "Design and Environment or How Political Economy Escalates into Cyberblitz," in *For a Critique of the Political Economy of the Sign,* trans. Charles Levin (St. Louis, Mo.: Telos Press, 1981), 185.

David Mitten | *The Axe Head*

1. Professor Cyril Stanley Smith of MIT was an innovator in showing how this could be done, in many cultures, in many media. See Cyril Stanley Smith, *From Art to Science* (Cambridge, Mass.: MIT Press, 1980).

Susan Spilecki | Dit Da Jow

1. See <http://www.castcarthtrade.com/ironpalm.php> (accessed on September 8, 2006).

Nathan Greenslit | *The Vacuum Cleaner*

1. Hans Selye, "Stress and Psychiatry," *The American Journal of Psychiatry* 113, no. 5 (1956): 423–427.
2. Karl Marx, *Capital: A Critique of Political Economy,* trans. Ben Fowkes (London: Penguin, 1976 [1867]), vol. I, 163.
3. Jean Baudrillard, *The System of Objects,* trans. James Benedict (London: Verso, 1996 [1968]).

William J. Mitchell | *The Melbourne Train*

1. Mark Twain, *Following the Equator* (Washington: National Geographic Adventure Classics, 2005 [1897]), 134.
2. Henry Lawson, "The Never-Never Land," in *Poetical Works of Henry Lawson* (Sydney: Angus and Robertson, 1984 [1906]), 113.
3. Alexander Pope, "An Essay on Criticism," in *The Poems of Alexander Pope* (London: Penguin, 1985 [1711]), 24.

Henry Jenkins | *Death-Defying Superheroes*

1. Umberto Eco, "The Myth of Superman," *Diacritics* 2, no. 1 (1972): 16.

Stefan Helmreich | *The SX-70 Instant Camera*

1. Victor K. McElheny, *Insisting on the Impossible: The Life of Edwin Land, Inventor of Instant Photography* (Cambridge, Mass.: Perseus Books, 1998), 358.
2. Howard G. Rogers, "Processes and Products for Forming Photographic Images in Color," US Patent # 2,983,606 (filed July 14, 1958; granted May 9, 1961).
3. McElheny, *Insisting,* 233.
4. Ibid., 221–222.

Susan Pollak | *The Rolling Pin*

1. Marcel Proust, *Remembrance of Things Past,* trans. C. K. Scott Moncrieff and Terence Kilmartin (New York: Vintage, 1981 [1913]), vol. 1, 48.
2. Ibid., 51.
3. D. W. Winnicott, "The Fate of the Transitional Object," in *Psychoanalytic Explorations,* ed. Clare Winnicott, Ray Shepherd, and Madeleine Davis (Cambridge, Mass.: Harvard University Press, 1989), 58.
4. Proust, *Remembrance,* vol. 3, 944.

Caroline A. Jones | *The Painting in the Attic*

1. D. W. Winnicott, "Transitional Objects and Transitional Phenomena" (1953) in *Playing and Reality* (New York: Routledge, 1989 [1971]), 10. I use the phrase "primary parent" to avoid the unhelpful fixation on the mother that attends most object-relations theories, including Winnicott's.
2. Ibid., 12.

Nancy Rosenblum | *Chinese Scholars' Rocks*

1. William Blake, "Auguries of Innocence," in *The Portable Romantic Poets,* ed. W. H. Auden and Norman Holmes Pearson (New York: Viking Penguin, 1978), 18.
2. Richard Rosenblum and Valerie Doran, *Art of the Natural World: Resonance of Wild Nature in Chinese Sculptural Art* (Boston: MFA Publications, 2001), 39.
3. <http:www.Rosenblumcollection.com> (accessed on January 23, 2007).

Susannah Mandel | *Apples*

1. Anthony Burgess, *A Clockwork Orange* (New York: W. W. Norton, 1987), 21–22.
2. Anthony Burgess, *A Clockwork Orange: A Play With Music,* based on the novella of the same name (London: Hutchinson, 1987), viii.
3. Lewis Thomas, *The Lives of a Cell: Notes of a Biology Watcher* (New York: Penguin, 1974), 45.

Jeffrey Mifflin | *The Mummy*

1. Blaise Pascal, *Pensées,* trans. A. J. Krailsheimer (London: Penguin, 1995 [1670]), 130.

Michael M. J. Fischer | *The Geoid*

1. A longer version of this essay, entitled "The Geoid as Transitional Object," is referenced on <http://web.mit.edu/anthropology/faculty_staff/fischer/publications.html>.

2. Irene K. Fischer, *Geodesy? What's That?: My Personal Involvement in the Age-Old Quest for the Size and Shape of the Earth* (Lincoln, Nebraska: iUniverse, 2005).
3. The Fischer Ellipsoid 1960, updated 1968, also known as Mercury Datum 1960 and modified Mercury Datum, 1968.

Robert P. Crease | *Foucault's Pendulum*

1. Umberto Eco, *Foucault's Pendulum,* trans. William Weaver (New York: Ballantine Books, 1990), 5.
2. Ibid., 201.

Evelyn Fox Keller | *Slime Mold*

1. Webster's Online Dictionary gives a rich description of slime mold:

> 1. An unusual fungus-like protist of the phylum Myxomycota or the class Myxomycetes, having a stage of growth in which it comprises a naked noncellular multinucleate mass of creeping protoplasm having characteristics of both plants and animals; it also has a propagative phase in which it develops fruiting bodies bearing spores; it is sometimes classified as a protist. It is called also *acellular slime mold.*
> 2. Any of several remarkable amoeba like organisms of the phylum Acrasiomycota, mostly terrestrial, having a fruiting phase resembling that of the acellular slime molds, but being cellular and nucleate throughout their life cycle; called also cellular slime mold. The most studied species is *Dictyostelium* discoideum. In their feeding phase, they live like amoebae as individual cells, engulfing bacteria as a prime food source. When the food source diminishes, they begin to aggregate, swarming together to form clumps which may move toward heat and light, so as to reach the surface of the ground; they then differentiate into

a form with spores contained within a sporangium resting on a stalk. When the spores are carried to another location with adequate food supplies, the spores may germinate to resume the life cycle. The phase of aggregation appears to be initiated by release of cyclic AMP, serving as a signal between the individual cells. The formation of the fruiting body has some similarities to differentiation in multicellular organisms, but the mechanisms are still under study. Some biologists object to the classification of *Dictyostelium* as a slime mold, as it is neither a mold nor slimy.

Available at <http://www.webster-dictionary.net/definition/slime%20mold> (accessed August 15, 2006).

2. Evelyn Fox Keller, *A Feeling for the Organism: The Life and Work of Barbara McClintock* (San Francisco: W. H. Freeman, 1983).

3. David L. Nanney, 1957, "The Role of the Cytoplasm in Heredity," in *The Chemical Basis of Heredity,* ed. W. D. McElroy and B. Glass (Baltimore: Johns Hopkins Press, 1957), 134–163.

4. See Evelyn Fox Keller, *Reflections on Gender and Science* (New Haven: Yale University Press, 1985), 157.

5. Closely related are arguments for the reframing of problems of development in terms of *Developmental Systems Theory,* a term that is itself sometimes (if somewhat mistakenly) used interchangeably with *Dynamical Systems Theory.* For a more extensive discussion of the various uses of the terms *Developmental* and *Dynamical Systems Theory,* see Evelyn Fox Keller, "DDS: Dynamics of Developmental Systems," *Biology and Philosophy* 20, nos. 2–3 (March 2005): 409–416.

6. *The Oxford English Dictionary,* 2nd ed. (Oxford: Clarendon Press, 1989), Vol. 12, 589.

7. See Evelyn Fox Keller, *The Century of the Gene* (Cambridge, Mass.: Harvard University Press, 2000), chapter 4, for further discussion of developmental robustness.

Sherry Turkle | *What Makes An Object Evocative?*

1. I thank my research assistant Anita Chan whose work on sources was both meticulous and brimming with good ideas. Ongoing conversations with Kelly Gray enhanced the clarity of this note.

2. As anthropologist Lucy Suchman puts it: "Objects and their positions are inseparable." See "Affiliative Objects," *Organization* 12, no. 3 (2005): 379–399. On this theme she cites Donna Haraway, *Simians, Cyborgs and Women: The Reinvention of Nature* (New York: Routledge, 1991); Andrew Pickering, *The Mangle of Practice: Time, Agency, and Science* (Chicago: University of Chicago Press, 1997); Marilyn Strathern, *Property, Substance, and Effect: Anthropological Essays on Persons and Things* (London: Athlone Press, 1999), and Karen Barad, "Posthumanist Performativity: Toward an Understanding of How Matter Comes to Matter," *Signs: Journal of Women in Culture and Society* 28, no.3 (2003): 88–128.

3. William Carlos Williams, *Paterson* (New York: New Directions, 1946), Book I, 7.

4. Claude Lévi-Strauss, *The Savage Mind,* trans. John Weightman and Doreen Weightman (Chicago: University of Chicago Press, 1966), 24.

5. Resnick's experience of the pull-toy as an evocative object is close to what the sociologist Karin Knorr Cetina calls an "epistemic object." Knorr Cetina describes epistemic objects as open, question-generating and complex: "They are processes and projections rather than definitive things. . . . Objects of knowledge . . . are more like open drawers filled with folders extending indefinitely into the depths of a dark closet." From "Sociality with Objects: Social Relations in Postsocial Knowledge Societies," *Theory, Culture, and Society* 14, no. 4 (1997): 12.

6. We form our notions of these things through immersion in our culturally specific object world. And when that culture changes, when it offers new objects, we can come to see the world differently. This notion of objects and culture

change is central to Resnick's work with StarLogo, an effort to provide children with a new world of objects—in this case computational objects—that will enable thinking about emergence. See Mitchel Resnick, *Turtles, Termites, and Traffic Jams: Explorations in Massively Parallel Microworlds* (Cambridge, Mass.: MIT Press, 1994). Seymour Papert makes this point about the power of computer programming as a microworld for learning in *Mindstorms: Children, Computers, and Powerful Ideas* (New York: Basic Books, 1980). While Piaget was a constructivist, stressing how objects are part of children's building their minds, Papert and Resnick consider themselves constructionists, putting the emphasis on children building for themselves the materials that will structure their thinking. For an overview of this activist, "builder's" perspective on objects and intellect, see Idit Harel and Seymour Papert, eds., *Constructionism: Research Reports and Essays, 1985–90, by the MIT Epistemology and Learning Group, the MIT Media Laboratory* (Norwood, N.J.: Ablex, 1991).

7. See Papert, *Mindstorms*, viii.

8. The psychoanalyst Jacques Lacan had a particular interest in the practice of knotting and how it can integrate the epistemic and the emotional. On this point, see Sherry Turkle, *Psychoanalytic Politics: Jacques Lacan and Freud's French Revolution* (Guilford, Conn.: Guilford Press, 1991 [1978]).

9. D. W. Winnicott, *Playing and Reality* (London: Routledge, 1989 [1971]), 15–20.

10. Lewis Mumford, *Technics and Civilization* (New York: Harcourt, Brace & World, 1963 [1934]).

11. Michel Foucault, *Discipline and Punish: The Birth of the Prison,* trans. Alan Sheridan (New York: Pantheon Books, 1977); *The Birth of the Clinic: An Archeology of Medical Perception,* trans. A. M. Sheridan Smith (New York: Vintage, 1994 [1963]); *Madness and Civilization: A History of Insanity in the Age of Reason,* trans. Richard Howard (New York: Random House, 1965).

12. The literary theorist Roland Barthes describes the objects of disciplinary society as "self-naturalizing." We begin to accept what we have made as what has always been and what must always be. Roland Barthes, *Mythologies,* trans. Annette Lavers (New York: Hill and Wang, 1972 [1957]).

13. Literary theorists Julia Kristeva and Judith Butler approach similar issues from a psychoanalytic perspective, influenced by the work of the psychoanalyst Jacques Lacan who elaborates on language as constitutive of self. Butler stresses that objects, including the body, are never found "in the raw." They are already shaped by language and social discourse and thus ready to shape us. For Julia Kristeva, each individual's psychology reproduces the political relationships of the outer world. So, any change we wish to make in the world first requires a parallel action within. Foucault describes an inner history of objects as it plays out on a social level; Kristeva and Butler bring the story down to the psychodynamics of individuals. See Judith Butler, *Bodies That Matter: On the Discursive Limits of "Sex"* (New York: Routledge, 1993); and Julia Kristeva, *Strangers to Ourselves,* trans. Leon Roudiez (New York: Columbia University Press, 1991).

14. The antidepressants become what sociologist and historian Bruno Latour would call the "foot soldiers" of medical power. The pills confirm individuals in the role of patient. See Bruno Latour, *The Pasteurization of France,* trans. Alan Sheridan and John Law (Cambridge, Mass.: Harvard University Press, 1988).

15. See Butler, *Bodies That Matter.*

16. Karl Marx, *Capital: A Critique of Political Economy,* trans. Ben Fowkes (London: Penguin, 1976 [1867]), vol. I, 163.

17. Marcel Mauss, *The Gift: The Form and Reason for Exchange in Archaic Societies,* trans. W. D. Halls (New York: W. W. Norton, 2000 [1950]), 11–12.

18. Jean Baudrillard, *For A Critique of the Political Economy of the Sign,* trans. Charles Levin, (St. Louis: Telos Press, 1981).

19. In *We Have Never Been Modern* Bruno Latour describes the subversion of objects as they reveal the cracks in the social constructions that we take as "reality," for example what we call "modernity." We have constructed a notion of modernity that rests on dichotomies between past and present, natural and social, and humans and things, all distinctions that our ancestors, in their world of alchemy, astrology, and phrenology, did not make. Our objects challenge these distinctions. Objects insist on their place in history, on the fact that they are embedded in both nature and culture, and finally, the objects of modern science refuse the distinction between humans and things. On a global level, objects such as "the ozone layer" fuse science, nature, technology, and politics. Technoscientific objects make manifest the social relations of our time. See Bruno Latour, *We Have Never Been Modern,* trans. Catherine Porter (Cambridge, Mass.: Harvard University Press, 1993). See also Bruno Latour and Steven Woolgar, *Laboratory Life: The Social Construction of Scientific Facts* (Princeton, N.J.: Princeton University Press, 1986 [1979]; Bruno Latour, *Science in Action: How to Follow Scientists and Engineers through Society* (Cambridge, Mass.: Harvard University Press, 1987); *The Pasteurization of France,* trans. Alan Sheridan and John Law (Cambridge, Mass.: Harvard University Press, 1988); *Aramis or the Love of Technology,* trans. Catherine Porter (Cambridge, Mass.: Harvard University Press, 1996).

20. For Winnicott, "This area of playing is not inner psychic reality. It is outside the individual but it is not the external world. Into this play area the child gathers objects or phenomena from external reality and uses these in the service of some sample derived from inner or personal reality." Winnicott, *Playing and Reality,* 51.

21. Victor Turner, *The Ritual Process: Structure and Anti-Structure* (Chicago: Aldine, 1969). Turner drew his notion of the liminal from the work of Arnold van Gennep. See

Rites of Passage, trans. Monika B. Vizedom and Gabrielle L. Caffee (Chicago: University of Chicago Press, 1960 [1909]).

22. Roland Barthes, *The Pleasure of the Text,* trans. Richard Miller (New York: Farrar, Straus and Giroux, 1975), 67.

23. Jacques Lacan, "On the Signification of the Phallus," in *Écrits: A Selection,* trans. Alan Sheridan (New York: W. W. Norton, 1977). For Lacan, the Oedipal period is the time when language and society come to inhabit the child, providing entrée into the "symbolic" order. In Lacan's telling, the Oedipal crisis is resolved through language. The child accepts the father's interdiction to the child's desire for the mother by identifying with the father through an acceptance of his name. (In French there is a pun: the father's *non* and *nom.*) One signifier (the father's name) comes to take the place of another (desire for the mother and desire to be the object of her desire). What is being signified (the primitive, irreducible desire to complete the mother, to be what she is presumed to lack) remains the same. In the course of a lifetime, the individual builds up many chains of signification, always substituting new terms for old and always increasing the distance between the signifier that is most accessible, and all those that are invisible and unconscious, including of course the original signifier. See Lacan, "On a Question Preliminary to any Possible Treatment of Psychosis," in *Écrits,* 200. See also Lacan, "The Function and Field of Speech and Language in Psychoanalysis," in *Écrits,* 68.

24. Joan Didion, *The Year of Magical Thinking* (New York: Knopf, 2005).

25. See Sigmund Freud, "Mourning and Melancholia," *The Standard Edition of the Complete Psychological Works of Sigmund Freud,* trans. and ed. James Strachey et al. (London: Hogarth Press, 1953–1974), vol. XIV, 239–258. Loss and what follows are central to the way psychoanalytic theory tells the story of the evolving self. When Lacan describes the Oedipal moment as the internalization of language and social law, his narrative parallels that of

Sigmund Freud. Freud sketched out the process through which loss becomes the motor for objects (here referring to people) to become part of our inner world. Freud considers the moment when a male child gives up the idea that he will displace his father in his mother's affections. The loss requires a radical response. The father, or rather the father's interdiction, is brought inside by creating a mental agency capable of representing it: the superego. (This is the moment that Lacan theorized as the moment of passage to a symbolic order.) Later losses will not result in new psychic structures at the level of the superego. But the basic process has been laid down: a loss, an internalization of an object. Mourning is the painful but necessary process through which this internalization takes place.

26. Psychologists Mihaly Csikszentmihalyi and Eugene Rochberg-Halton investigated people's feelings toward household objects such as the rolling pin, finding that they carry history and meaning. See Mihaly Csikszentmihalyi and Eugene Rochberg-Halton, *The Meaning of Things: Domestic Symbols and the Self* (Cambridge: Cambridge University Press, 1981).

27. Marcel Proust, *Remembrance of Things Past,* trans. C. K. Scott Montcrieff and Terence Kilmartin (New York: Vintage, 1981 [1913]), vol. 1, 48–51.

28. Mr. B's evokes Henri Bergson as well as Proust. Bergson criticizes any description of memory that divides it into a series of mechanical steps removed from sensation. For Bergson, we must physically prepare ourselves to recall. Remembering is not to re-experience the same thing, but to experience something new and to draw the past into a new realm of possibility. See Henri Bergson, *Matter and Memory*, trans. Nancy M. Paul and W. Scott Palmer (New York: Zone Books, 1990 [1896]), 133–135.

29. Proust, *Remembrance of Things Past,* vol. 3, 994.

30. The rocks carry ideas and inspire passion. They present thought and feeling as inseparable. In microcosm, the rocks capture what Immanuel Kant called nature's

"sublime" and illustrate Gaston Bachelard's thesis that only the infusion of reason with passion can lead to a full comprehension of the world. Immanuel Kant, *The Critique of Judgment,* trans. James C. Meredith (Oxford: Oxford University Press, 1952 [1790]); Gaston Bachelard, *The Poetics of Space: The Classic Look at How We Experience Intimate Places,* trans. Maria Jolas (Boston: Beacon Press, 1994 [1958]).

31. Sigmund Freud, "The Uncanny," *The Standard Edition of the Complete Psychological Works of Sigmund Freud,* trans. and ed. James Strachey et al., vol. XVII, 219–252.

32. Like other boundary objects, dirt and pollution provoke strong emotions and carry powerful ideas. To make this point, Mary Douglas looks at the taboo foods of Jewish dietary tradition. In Douglas's formulation, every kosher meal embodies the ordered cosmology of Genesis in which heaven was separated from the earth and the sea. In the story of the creation, each of these realms is alotted its proper kind of animal life. In the heavens, two-legged fowls fly with wings; on the earth, four-legged animals hop or walk; and scaly fish swim with fins. It is acceptable to eat these "pure" creatures who behave in harmony with their realm, but those that cross categories (such as the lobster that lives in the sea but crawls upon its floor) are unacceptable. Every kosher meal carries a theory of unbreachable order. Holiness is order and each thing must have its place. Mary Douglas, *Purity and Danger: An Analysis of the Concepts of Pollution and Taboo* (New York: Routledge, 1966), especially chapter 3, "The Abominations of Leviticus."

33. Evelyn Fox Keller, *Reflections on Gender and Science* (New Haven, Conn.: Yale University Press, 1985), 157.

34. Jacques Derrida, *Archive Fever: A Freudian Impression,* trans. Eric Prenowitz (Chicago: University of Chicago Press, 1996), 17–18.

35. Jacques Derrida, "The Word Processor," in *Paper Machine,* trans. Rachel Bowlby (Stanford: Stanford University Press, 2005), 24

36. Donna J. Haraway, "A Cyborg Manifesto," in *Simians, Cyborgs, and Women: The Reinvention of Nature.* (New York: Routledge, 1991), 162.

37. Haraway, "A Cyborg Manifesto," in *Simians, Cyborgs, and Women,* 162.

38. Ibid., 163

Selected Bibliography

Akrich, Madeleine. "The Description of Technological Objects." In *Shaping Technology/Building Society*, edited by W. E. Bijker and J. Law. Cambridge, Mass.: MIT Press, 1994.

Althusser, Louis. *Essays on Ideology*. Translated by Ben Brewster et al. London: Verso, 1984.

Andersen, Hans Christian. *Fairy Tales*, edited by Jackie Wullschlager. Translated by Tiina Nunnally. New York: Viking, 2005 [1845].

Appadurai, Arjun, ed. *The Social Life of Things: Commodities in Cultural Perspective*. Cambridge: Cambridge University Press, 1988.

Bachelard, Gaston. *The Poetics of Space: The Classic Look at How We Experience Intimate Places*. Translated by Maria Jolas. Boston: Beacon Press, 1994 [1958].

Bakhtin, M. M. *The Dialogic Imagination: Four Essays*, edited by Michael Holquist. Translated by Michael Holquist and Caryl Emerson. Austin: University of Texas Press, 1981.

Balsamo, Anne. *Technologies of the Gendered Body: Reading Cyborg Women*. Durham, N.C.: Duke University Press, 1996.

Barad, Karen. "Posthumanist Performativity: Toward an Understanding of How Matter Comes to Matter." *Signs: Journal of Women in Culture and Society* 28, no. 3 (2003): 88–128.

Barry, Andrew. "On Interactivity: Consumers, Citizens, and Culture." In *The Politics of Display: Museum,*

Science, Culture, edited by Sharon Macdonald. London: Routledge, 1998.

Barthes, Roland. *Empire of Signs.* Translated by Richard Howard. New York: Hill and Wang, 1982.

—. *Mythologies.* Translated by Annette Lavers. New York: Hill and Wang. 1972 [1957].

—. *The Pleasure of the Text.* Translated by Richard Miller. New York: Farrar, Straus and Giroux, 1975.

Baudrillard, Jean. *For a Critique of the Political Economy of the Sign.* Translated by Charles Levin. St. Louis, Mo.: Telos Press, 1981.

—. *Simulacra and Simulations.* Translated by Sheila Faria Glaser. Ann Arbor: University of Michigan Press, 1995.

—. *The System of Objects.* Translated by James Benedict. London: Verso, 1996 [1968]).

Beniger, James R. *The Control Revolution: Technological and Economic Origins of the Information Society.* Cambridge, Mass.: Harvard University Press, 1986.

Benjamin, Walter. *Illuminations,* edited with introduction by Hannah Arendt. Translated by Harry Zohn. New York: Harcourt, Brace & World, 1968.

Berger, Peter L., and Thomas Luckmann. *The Social Construction of Reality: A Treatise in the Sociology of Knowledge.* Garden City, N.Y.: Doubleday, 1966.

Bergson, Henri. *Matter and Memory.* Translated by Nancy M. Paul and W. Scott Palmer. New York: Zone Books, 1990 [1896].

Bhabha, Homi K. *The Location of Culture.* London: Routledge, 1994.

Bijker, Wiebe E. *Bicycles, Bakelites, and Bulbs: Toward a Theory of Sociotechnical Change.* Cambridge, Mass.: MIT Press, 1995.

Bijker, Wiebe E., Thomas P. Hughes, and Trevor J. Pinch, eds. *The Social Construction of Technological Systems: New Directions in the Sociology and History of Technology.* Cambridge, Mass.: MIT Press, 1987.

Blanchot, Maurice. *Awaiting Oblivion.* Translated by John Gregg. Lincoln: University of Nebraska Press, 1997.

Bleier, Ruth, ed. *Feminist Approaches to Science.* New York: Pergamon Press, 1986.

Bollas, Christopher. *The Shadow of the Object: Psychoanalysis of the Unthought Known.* London: Free Association Books, 1987.

Bourdieu, Pierre. *Distinction: A Social Critique of the Judgement of Taste.* Translated by Richard Nice. London: Routledge, 1984.

————. *The Logic of Practice.* Translated by Richard Nice. Stanford, Calif.: Stanford University Press, 1990.

————. *Outline of a Theory of Practice.* Translated by Richard Nice. Cambridge: Cambridge University Press, 1977.

Breton, André. "Surrealist Situation of the Object." In *Manifestos of Surrealism.* Ann Arbor: University of Michigan Press, 1972.

Brosterman, Norman. *Inventing Kindergarten.* New York: Harry N. Abrams, 1997.

Brown, Bill, ed. *Things.* Chicago: University of Chicago Press, 2004.

Bukatman, Scott. *Terminal Identity: The Virtual Subject in Postmodern Science Fiction.* Durham, N.C.: Duke University Press, 1993.

Burke, Edmund. *A Philosophical Enquiry into the Origin of Our Ideas of the Sublime and Beautiful.* Oxford: Oxford University Press, 1990 [1757].

Butler, Judith. *Bodies That Matter: On the Discursive Limits of "Sex."* New York: Routledge, 1993.

———. *Gender Trouble: Feminism and the Subversion of Identity (Thinking Gender).* New York: Routledge, 1989.

Clynes, Manfred, and Nathan Kline. "Cyborgs and Space." *Astronautics* 14 (1960): 26–27, 74–75.

Crease, Robert P. *The Prism and the Pendulum: The Ten Most Beautiful Experiments in Science.* New York: Random House, 1993.

Csikszentmihalyi, Mihaly, and Eugene Rochberg-Halton. *The Meaning of Things: Domestic Symbols and the Self.* Cambridge: Cambridge University Press, 1981.

Csordas, Thomas J. "Embodiment as a Paradigm for Anthropology." *Ethos* 18, no. 1 (1990): 5–47.

Daston, Lorraine, ed. *Biographies of Scientific Objects.* Chicago: University of Chicago Press, 2000.

———. *Things That Talk: Object Lessons from Art and Science.* New York: Zone Books, 2004.

Deleuze, Gilles, and Félix Guattari. *Anti-Oedipus: Capitalism and Schizophrenia.* Translated by Robert Hurley, Mark Seem, and Helen Lane. New York: Viking Press, 1977.

———. *A Thousand Plateaus: Capitalism and Schizophrenia.* Translated by Brian Massumi. Minneapolis: University of Minnesota Press, 1987.

Derrida, Jacques. *Archive Fever: A Freudian Impression.* Translated by Eric Prenowitz. Chicago: University of Chicago Press, 1996.

———. *The Gift of Death, Religion and Postmodernism.* Translated by David Wills. Chicago: University of Chicago Press, 1995.

———. *Paper Machine.* Translated by Rachel Bowlby. Stanford, Calif.: Stanford University Press, 2005.

———. *Postcard: From Socrates to Freud and Beyond.* Translated by Alan Bass. Chicago: University of Chicago Press, 1987.

———. *Specters of Marx: The State of the Debt, the Work of Mourning, and the New International.* Translated by Peggy Kamuf. New York: Routledge, 1994.

Didion, Joan. *The Year of Magical Thinking.* New York: Knopf, 2005.

Dillard, Annie. *Teaching a Stone to Talk: Expeditions and Encounters.* New York: Harper and Row, 1982.

Douglas, Mary. *Purity and Danger: An Analysis of Concepts of Pollution and Taboo.* New York: Routledge, 1992 [1966].

Douglas, Mary, and B. Isherwood. *The World of Goods: Towards an Anthropology of Consumption.* London: Routledge, 1996.

Durkheim, Emile. *The Division of Labor in Society.* Translated by W. D. Halls. New York: Free Press, 1984 [1893].

———. *The Elementary Forms of the Religious Life.* Translated by Karen E. Fields. New York: Free Press, 1995 [1912].

———. *The Rules of Sociological Method.* Translated by W. D. Halls. New York: Free Press, 1982 [18895]).

———. *Suicide, a Study in Sociology,* edited by George Simpson. Translated by John A. Spaulding and George Simpson. Glencoe, Ill.: Free Press, 1951 [1897].

Durkheim, Emile, and Marcel Mauss. *Primitive Classification*. Translated by Rodney Needham. Chicago: University of Chicago Press, 1963 [1903].

Eco, Umberto. *Foucault's Pendulum*. Translated by William Weaver. New York: Ballantine Books, 1990.

————. "The Myth of Superman." *Diacritics* 2, no. 1 (1972): 14–22.

————. *Travels in Hyperreality*. San Diego: Harcourt Brace Jovanovich, 1986.

Engels, Friedrich. "The Origin of Family, Private Property, and the State." In *The Marx-Engels Reader,* edited by Robert C. Tucker. New York: W. W. Norton, 1978 [1884].

Erikson, Erik. *Childhood and Society*. New York: W. W. Norton, 1963.

Evans-Pritchard, E. E. *Witchcraft, Oracles, and Magic among the Azande*. Oxford: Clarendon Press, 1937.

Feynman, Richard. *Surely You're Joking, Mr. Feynman*. New York: W. W. Norton, 1981.

Flusser, Vilém. "Digital Apparition." In *Electronic Culture: Technology and Visual Representation,* edited by Timothy Druckrey. New York: Aperture, 1996.

Foucault, Michel. *The Birth of the Clinic: An Archeology of Medical Perception*. 1963. Translated by A. M. Sheridan Smith. New York: Vintage, 1994 [1963].

————. *Discipline and Punish: The Birth of the Prison*. Translated by Alan Sheridan. New York: Pantheon Books, 1977.

————. *The History of Sexuality*. Translated by Robert Hurley. New York: Pantheon Books, 1978.

———. *Madness and Civilization: A History of Insanity in the Age of Reason.* Translated by Richard Howard. New York: Random House, 1965.

———. "Truth and Power." In *Power/Knowledge: Selected Interviews and Other Writings, 1972–1977,* edited by Colin Gordon. Translated by Colin Gordon et al. New York: Pantheon Books, 1980.

Freud, Sigmund. *The Standard Edition of the Complete Works of Sigmund Freud.* Translated and edited by James Strachey et al. London: Hogarth Press, 1953–1974. See esp. "Fetishism," "The Interpretation of Dreams," "Mourning and Melancholia." "Parapraxes," and "The Uncanny."

Froebel, Friedrich. *The Education of Man.* Translated and annotated by W. N. Hailman. Mineola, N.Y.: Dover Publications, 2005 [1892].

Galison, Peter. "Aufbau/Bauhaus: Logical Positivism and Architectural Modernism." *Critical Inquiry* 16, no. 4 (1990): 709–752.

———. *Image and Logic: A Material Culture of Microphysics.* Chicago: University of Chicago Press, 1997.

Garfinkel, H. *Studies in Ethnomethodology.* New York: Prentice-Hall, 1967.

Geertz, Clifford. *The Interpretation of Cultures: Selected Essays.* New York: Basic Books, 1973.

Giddens, Anthony. *New Rules of Sociological Method: A Positive Critique of Interpretative Sociologies.* New York: Basic Books, 1976.

Gieryn, Thomas. "Balancing Acts: Science, Enola Gay, and History Wars at the Smithsonian." In *The Politics of Display: Museums, Science, Culture,* edited by Sharon Macdonald. London: Routledge, 1998.

Gilligan, Carol. *In a Different Voice: Psychological Theory and Women's Development.* Cambridge, Mass., Harvard University Press, 1982.

Godelier, Maurice. *The Enigma of the Gift.* Translated by Nora Scott. Chicago: University of Chicago Press, 1999.

Goffman, Erving. *The Presentation of Self in Everyday Life.* Garden City, N.Y.: Doubleday, Anchor Books, 1959.

Goldman, Lucien. *Cultural Creation in Modern Society.* Translated by Bart Grahl. Oxford: B. Blackwell, 1977.

Goodman, Nelson. *Ways of Worldmaking.* Indianapolis: Hackett, 1978.

Gould, Glenn. "The Prospects of Recording." In *The Glenn Gould Reader,* edited by Tim Page. New York: Vintage, 1984.

Gramsci, Antonio. *Selections from the Prison Notebooks,* edited and translated by Quintin Hoare and Geoffrey Nowell-Smith. London: Lawrence and Wishart, 1971.

Grosz, Elizabeth A. *Volatile Bodies: Toward A Corporeal Feminism.* Bloomington: Indiana University Press, 1994.

Haraway, Donna J. "The Cyborg Manifesto." In *Simians, Cyborgs, and Women: The Reinvention of Nature.* New York: Routledge, 1991.

———. *Modest-Witness@Second-Millennium.Femaleman-Meets-Oncomouse: Feminism and Technoscience.* New York: Routledge, 1997.

———. "Postscript." In *Technoculture,* edited by Constance Penley and Andrew Ross. Minneapolis: University of Minnesota Press, 1991.

Harding, Sandra G., ed. *The Feminist Standpoint Theory Reader: Intellectual and Political Controversies.* New York: Routledge, 2004.

Harding, Sandra, and Merrill B. Hintikka, eds. *Discovering Reality: Feminist Perspectives on Epistemology, Metaphysics, Methodology, and Philosophy of Science.* London: Reidel, 1983.

Harel, Idit, and Seymour Papert, eds. *Constructionism: Research Reports and Essays, 1985–90, by the MIT Epistemology and Learning Group, the MIT Media Laboratory.* Norwood, N.J.: Ablex, 1991.

Harre, Rom. "Material Objects in Social Worlds." *Theory, Culture and Society,* no. 19 (2002): 23–34.

Hayles, N. Katherine. *How We Became Posthuman: Virtual Bodies in Cybernetics, Literature and Informatics.* Chicago: University of Chicago Press, 1999.

————. "Simulated Nature and Natural Simulations: Rethinking the Relations between the Beholder and the World." In *Uncommon Ground,* edited by W. Cronin. New York: W. W. Norton, 1995.

Heidegger, Martin. *The Question Concerning Technology and Other Essays.* Translated by William Lovitt. New York: Harper and Row, 1977.

————. *What Is a Thing?* Translated by W. B. Barton Jr. and Vern Deutsch. Chicago: H. Regnery Co., 1968.

Henare, Amina, ed. *Thinking Through Things: Theorising Artifacts in Ethnographic Perspective.* London: UCL Press, 2006.

Horkheimer, Max, and Theodor W. Adorno. *Dialectic of Enlightenment.* Translated by Edmund Jephcott. Edited by Gunzelin Schmid Noerr. New York: Herder and Herder, 1972.

James, William. *The Varieties of Religious Experience.* New York: Signet, 2003 [1902].

Jameson, Fredric. *Postmodernism, or, The Cultural Logic of Late Capitalism.* Durham, N.C.: Duke University Press, 1991.

Johnson, Mark. *The Body in the Mind: The Bodily Basis of Meaning, Imagination and Reason.* Chicago: University of Chicago Press, 1987.

Jones, Caroline A., and Peter Galison. *Picturing Science: Producing Art.* New York: Routledge, 1998.

Kant, Immanuel. *The Critique of Judgment.* Translated by James C. Meredith. Oxford: Oxford University Press, 1952 [1790].

Keller, Evelyn Fox. *A Feeling for the Organism: The Life and Work of Barbara McClintock.* San Francisco: W. H. Freeman, 1983.

———. *Reflections on Gender and Science.* New Haven: Yale University Press, 1985.

Klein, Melanie. *Love, Guilt, and Reparation and Other Works, 1921–1945.* New York: Dell, 1975.

Knorr Cetina, Karin. *Epistemic Cultures: How the Sciences Make Knowledge.* Cambridge, Mass.: Harvard University Press, 1999.

———. *The Manufacture of Knowledge: An Essay on the Constructivist and Contextual Nature of Science.* Oxford: Pergamon Press, 1981.

———. "Sociality with Objects: Social Relations in Postsocial Knowledge Societies." *Theory, Culture, and Society* 14 (1997): 1–30.

Kopytoff, Igor. "The Cultural Biography of Things: Commoditization as Process." In *The Social Life of Things,* edited by Arjun Appadurai. Cambridge: Cambridge University Press, 1986.

Krips, Henry. *Fetish: An Erotics of Culture*. Ithaca, N.Y.: Cornell University Press, 1995.

Kristeva, Julia. *Powers of Horror: An Essay on Abjection, European Perspectives*. Translated by Leon Roudiez. New York: Columbia University Press, 1982.

———. *Strangers to Ourselves*. Translated by Leon Roudiez. New York: Columbia University Press, 1991.

Kuhn, Thomas S. *The Structure of Scientific Revolutions*. 2nd ed. Chicago: University of Chicago Press, 1970.

Lacan, Jacques. *Écrits: A Selection*. Translated by Alan Sheridan. New York: W. W. Norton, 1977.

———. *The Four Fundamental Concepts of Psycho-Analysis*. Translated by Alan Sheridan. New York: W. W. Norton, 1978.

Lakoff, George. *Women, Fire, and Dangerous Things: What Categories Reveal about the Mind*. Chicago: University of Chicago Press, 1987.

Lakoff, George, and Mark Johnson. *Philosophy in the Flesh: The Embodied Mind and Its Challenge to Western Philosophy*. New York: Basic Books, 1999.

Landow, George. *Hypertext 2.0: The Convergence of Contemporary Critical Theory and Technology*. Baltimore: Johns Hopkins University Press, 1997.

Latour, Bruno. *Aramis, or the Love of Technology*. Translated by Catherine Porter. Cambridge, Mass.: Harvard University Press, 1996.

———. *Pandora's Hope: Essays on the Reality of Science Studies*. Cambridge, Mass.: Harvard University Press, 1999.

———. *The Pasteurization of France*. Translated by Alan Sheridan and John Law. Cambridge, Mass.: Harvard University Press, 1988.

———. *Science in Action: How to Follow Scientists and Engineers through Society.* Cambridge, Mass.: Harvard University Press, 1987.

———. *We Have Never Been Modern.* Translated by Catherine Porter. Cambridge, Mass.: Harvard University Press, 1993.

Latour, Bruno, and Steven Woolgar. *Laboratory Life: The Social Construction of Scientific Facts.* Princeton, N.J.: Princeton University Press, 1986 [1979].

Lave, Jean. *Cognition in Practice: Mind, Mathematics, and Culture in Everyday Life.* Cambridge: Cambridge University Press, 1988.

Lavine, Steven, and Ivan Karp. *Exhibiting Cultures: The Poetics and Politics of Museum Display.* Washington, D.C.: Smithsonian Institution Press, 1991.

Leuenberger, Christine. "Constructing the Berlin Wall: How Material Culture Is Used in Psychological Theory." *Social Problems* 53, no. 1: 18–37.

Lévi-Strauss, Claude. *Structural Anthropology.* Translated by Monique Layton. New York: Basic Books, 1976.

———. *The Savage Mind.* Translated by John Weightman and Doreen Weightman. Chicago: University of Chicago Press, 1966.

———. *Totemism.* Translated by Rodney Needham. Boston: Beacon Press, 1963.

———. *Tristes Tropiques.* Translated by John Weightman and Doreen Weightman. New York: Atheneum, 1974.

Lubar, Steven, and David Kingery. *History from Things: Essays on Material Culture.* Washington, D.C.: Smithsonian Press, 1995.

Lyotard, Jean-François. *The Postmodern Condition: A Report on Knowledge, Theory and History of Literature,* Vol. 10. Minneapolis: University of Minnesota Press, 1984.

Malinowski, Bronislaw. *Argonauts of the Western Pacific: An Account of Native Enterprise and Adventure in the Archipelagoes of Melanesian New Guinea.* London: G. Routledge and Sons, 1922.

Mann, David W. *A Simple Theory of the Self.* New York: W. W. Norton, 1994.

Mannheim, Karl. *Essays on the Sociology of Culture.* New York: Oxford University Press, 1956.

Marx, Karl. *Capital: A Critique of Political Economy.* Translated by Ben Fowkes. London: Penguin, 1976 [1867].

————. *The Grundrisse,* edited and translated by David McLellan. New York: Harper and Row, 1971 [1858].

Mauss, Marcel. *The Gift: The Form and Reason for Exchange in Archaic Societies.* Translated by W. D. Halls. New York: W. W. Norton, 2000 [1950].

McLuhan, Marshall. *Understanding Media: The Extensions of Man.* New York: American Library, 1964.

Mead, Margaret. *Coming of Age in Samoa: A Psychological Study of Primitive Youth for Western Civilization.* New York: Blue Ribbon Books, 1928.

Merleau-Ponty, Maurice. *Phenomenology of Perception.* Translated by Colin Smith. London: Routledge, 1962.

Miller, Daniel. *Material Cultures: Why Some Things Matter.* Chicago: University of Chicago Press, 1998.

Mumford, Lewis. *Technics and Civilization.* New York: Harcourt, Brace & World, 1963 [1934].

Negroponte, Nicholas. *Being Digital.* New York: Vintage, 1996.

Norman, Donald A. *The Design of Everyday Things.* New York: Doubleday, 1990.

———. *Emotional Design: Why We Love (Or Hate) Everyday Things.* New York: Basic Books, 2004.

Papert, Seymour. *Mindstorms: Children, Computers, and Powerful Ideas.* New York: Basic Books, 1980.

Pasztory, Esther. *Thinking with Things.* Austin: University of Texas, 2005.

Penley, Constance, and Andrew Ross, eds. *Technoculture.* Minneapolis: University of Minnesota Press, 1991.

Petroski, Henry. *The Evolution of Useful Things.* New York: Knopf, 1993.

———. *The Pencil: A History of Design and Circumstance.* New York: Knopf, 1989.

Piaget, Jean. *The Child's Conception of the World.* 1929. Translated by Joan and Andrew Tomlinson. Totowa, N.J.: Littlefield Adams, 1960.

———. *Genetic Epistemology.* Translated by Eleanor Duckworth. New York: Columbia University Press, 1970.

Piaget, Jean, and Barbel Inhelder. *The Growth of Logical Thinking from Childhood to Adolescence.* Translated by Anne Parsons and Stanley Milgram. New York: Basic Books, 1958.

Pickering, Andrew. *The Mangle of Practice: Time, Agency, and Science.* Chicago: University of Chicago Press, 1997.

Pinch, Trevor J., and Wiebe E. Bijker. "The Social Construction of Facts and Artefacts: Or How the

Sociology of Science and the Sociology of Technology Might Benefit Each Other." *Social Studies of Science* 14: 388–441.

Pinch Trevor, and Frank Trocco. *Analog Days: The Invention and Impact of the Moog Synthesizer.* Cambridge, Mass.: Harvard University Press, 2002.

Poster, Mark. *The Mode of Information: Poststructuralism and Social Context.* Chicago: University of Chicago Press, 1990.

Proust, Marcel. *Remembrance of Things Past.* Translated by C. K. Scott Montcrieff and Terence Kilmartin. New York: Vintage, 1981 [1913].

Rabinow, Paul. "Artificiality and Enlightenment: From Sociobiology to Biosociality." In *Incorporations,* edited by Jonathan Crary and Sanford Kwinter. New York: Zone Books, 1992.

Rabinow, Paul. *Making PCR: A Story of Biotechnology.* Chicago: University of Chicago Press, 1996.

Radcliffe-Brown, A. R. *Structure and Function in Primitive Society: Essays and Addresses.* New York: Free Press, 1965.

Ratey, John J. *A User's Guide to the Brain: Perception, Attention, and the Four Theaters of the Brain.* New York: Vintage, 2002.

Resnick, Mitchel. *Turtles, Termites, and Traffic Jams: Explorations in Massively Parallel Microworlds.* Cambridge, Mass.: MIT Press, 1994.

Rheinberger, Hans-Jörg. *Toward a History of Epistemic Things: Synthesizing Proteins in the Test Tube.* Stanford, Calif.: Stanford University Press, 1997.

Rilke, Rainer Maria. *Letters to a Young Poet.* Translated by Stephen Mitchell. New York: Vintage, 1986 [1906].

Romanyshyn, Robert. *Technology as Symptom and Dream.* London: Routledge, 1989.

Rosch, Eleanor, Evan Thompson, and Francisco Varela. *The Embodied Mind: Cognitive Science and Human Experience.* Cambridge, Mass.: MIT Press, 1992.

Rosenblum, Richard, and Valerie Doran. *Art of the Natural World: Resonances of Wild Nature in Chinese Sculptural Art.* Boston: MFA Publications, 2001.

Sacks, Oliver. *Uncle Tungsten: Memories of a Chemical Boyhood.* New York: Knopf, 2001.

Sahlins, Marshall David. *Culture and Practical Reason.* Chicago: University of Chicago Press, 1976.

Said, Edward W. *Orientalism.* New York: Pantheon Books, 1978.

Saussure, Ferdinand de. *Course in General Linguistics.* Translated by W. Baskin. New York: Philosophical Library, 1959.

Schön, Donald A. "Designing as Reflective Conversation with the Materials of a Design Situation." *Knowledge-Based Systems* 5, no. 1 (1992): 3–13.

Searle, John. "Minds, Brains, and Programs." In *Behavioral and Brain Sciences* 3 (1980): 417–457.

Serres, Michel, and Bruno Latour. *Conversations on Science, Culture, and Time: Studies in Literature and Science.* Translated by Roxanne Lapidus. Ann Arbor: University of Michigan Press, 1995.

Smith, Cyril Stanley. *From Art to Science.* Cambridge, Mass.: MIT Press, 1980.

Sontag, Susan. *On Photography.* New York: Dell, 1978.

Star, Susan Leigh, and Geoffrey C. Bowker. *Sorting Things Out: Classification and Its Consequences.* Cambridge, Mass.: MIT Press, 2000.

Star, Susan Leigh, and James R. Griesemer. "Institutional Ecology, 'Translations,' and Boundary Objects: Amateurs and Professionals in Berkeley's Museum of Vertebrate Zoology." *Social Studies of Science* 19 (1989): 387–420.

Strathern, Marilyn. *Property, Substance, and Effect: Anthropological Essays on Persons and Things.* London: Athlone Press, 1999.

Suchman, Lucy. "Affiliative Objects." *Organization* 12, no. 3 (2005): 379–399.

———. *Plans and Situated Actions: The Problem of Human-Machine Communication.* Cambridge: Cambridge University Press, 1987.

Tenner, Edward. *Our Own Devices: The Past and Future of Body Technology.* New York: Knopf, 2003.

———. *When Things Bite Back.* New York: Knopf, 1996.

Thomas, Lewis. *The Lives of a Cell: Notes of a Biology Watcher.* New York: Penguin, 1974.

Thomas, Nicholas. *Entangled Objects: Exchange, Material Culture, and Colonialism in the Pacific.* Cambridge, Mass.: Harvard University Press, 1991.

Tilley, Christopher, ed. *Reading Material Culture: Structuralism, Hermeneutics, and Post-Structuralism.* Oxford: B. Blackwell, 1991.

Traweek, Sharon. *Beamtimes and Lifetimes: The World of High Energy Physicists.* Cambridge, Mass.: Harvard University Press, 1988.

———. "Border Crossings: Narrative Strategies in Science Studies and among Physicists in Tsukuba Science City, Japan." In *Science as Practice and Culture,* edited by Andrew Pickering. Chicago: University of Chicago Press, 1992.

Turkle, Sherry. *Life on the Screen: Identity in the Age of the Internet.* New York: Simon and Schuster, 1995.

———. *The Second Self: Computers and the Human Spirit.* Cambridge, Mass.: MIT Press, 2005 [1984].

———. "Whither Psychoanalysis in Computer Culture." *Psychoanalytic Psychology: Journal of the Division of Psychoanalysis* 21, vol. 1 (2004): 16–30.

Turkle, Sherry, and Seymour Papert. "Epistemological Pluralism: Styles and Voices within the Computer Culture." *Signs: Journal of Women in Culture and Society* 16, no. 1 (1990): 128–157.

Turner, Victor. *The Forest of Symbols: Aspects of Ndembu Ritual.* Ithaca, N.Y.: Cornell University Press, 1967.

———. *The Ritual Process: Structure and Anti-Structure.* Chicago: Aldine, 1969.

van Gennep, Arnold. *Rites of Passage.* Translated by Monika B. Vizedom and Gabrielle L. Caffee. Chicago: University of Chicago Press, 1960 [1909].

Vygotsky, Lev. *Mind in Society,* edited by Michael Cole, Vera John-Steiner, Sylvia Scribner, and Ellen Souberman. Cambridge, Mass.: Harvard University Press, 1978.

Weber, Max. *Economy and Society: An Outline of Interpretive Sociology,* edited by Guenther Roth and Claus Wittich. Translated by Ephraim Fischoff et al. New York: Bedminster Press, 1968.

———. *The Protestant Ethic and the Spirit of Capitalism.* Translated by Talcott Parsons. New York: Routledge, 1992 [1904].

Weizenbaum, J. *Computer Power and Human Reason: From Judgment to Calculation.* New York: W. H. Freeman, 1979.

Whorf, Benjamin Lee. *Language, Thought, and Reality: Selected Writings,* edited by John Bissell Carroll. Cambridge, Mass.: MIT Press, 1956.

Wiener, Norbert. *God and Golem, Inc.: A Comment on Certain Points Where Cybernetics Impinges on Religion.* Cambridge, Mass.: MIT Press, 1966.

Williams, William Carlos. *Paterson.* New York: New Directions, 1946.

Winner, Langdon. "Do Artifacts Have Politics?" In *Daedalus* 109, vol. 1 (1981): 121–136.

Winnicott, D. W. *Collected Papers: Through Paediatrics to Psycho-Analysis.* London: Tavistock Publications; New York: Basic Books, 1958.

———. *Playing and Reality.* New York: Routledge, 1989 [1971].

———. *Psychoanalytic Explorations,* edited by Clare Winnicott, Ray Shepherd, and Madeleine Davis. Cambridge, Mass.: Harvard University Press, 1989.

———. "The Use of an Object." *International Journal of Psychoanalysis* 50 (1969): 711–716.

Winograd, Terry. "Heidegger and the Design of Computer Systems." In *Technology and the Politics of Knowledge,* edited by Andrew Feenberg and Alastair Hannay. Bloomington: Indiana University Press, 1995.

Woolgar, Steve, and Geoff Cooper. "Do Artefacts Have Ambivalence?" In *Social Studies of Science* 29, vol. 3 (1999): 433–449.

Žižek, Slavoj. "Cyberspace, or, the Unbearable Closure of Being." *Pretexts: Studies in Writing and Culture* 6, vol. 1 (1997): 53–79.

———. *Mapping Ideology.* London: Verso, 1994.

———. *The Puppet and the Dwarf: The Perverse Core of Christianity.* Cambridge, Mass.: MIT Press, 2003.

Epigraph Sources

Tod Machover | *My Cello*

Erik Erikson, *Childhood and Society* (New York: W. W. Norton, 1963), 222.

Carol Strohecker | *Knots*

Claude Lévi-Strauss, *The Savage Mind,* trans. John Weightman and Doreen Weightman (Chicago: University of Chicago Press, 1966), 17, 21.

Susan Yee | *The Archive*

Jacques Derrida, *Archive Fever,* trans. Eric Prenowitz (Chicago: University of Chicago Press, 1996), 17–18.

Mitchel Resnick | *Stars*

Jean Piaget. *Genetic Epistemology,* trans. Eleanor Duckworth (New York: Columbia University Press, 1970), 15–16.

Howard Gardner | *Keyboards*

Lev Vygotsky, *Mind in Society,* ed. Michael Cole, Vera John-Steiner, Sylvia Scribner, and Ellen Souberman (Cambridge, Mass.: Harvard University Press, 1978), 92–104.

Eden Medina | *Ballet Slippers*

Hans Christian Anderson, "The Red Shoes," in *Fairy Tales,* trans. Tiina Nunnally, ed. Jackie Wullschlager (New York: Viking, 2005), 207–209.

Joseph Cevetello | *The Elite Glucometer*

Donna J. Haraway, "The Cyborg Manifesto," in *Simians, Cyborgs, and Women: The Reinvention of Nature* (New York: Routledge, 1991), 162–163.

Matthew Belmonte | *The Yellow Raincoat*

Julia Kristeva, *Strangers to Ourselves,* trans. Leon Roudiez (New York: Columbia University Press, 1991), 187, 191.

Michelle Hlubinka | *The Datebook*

Lewis Mumford, *Technics and Civilization* (New York: Harcourt, Brace & World, 1963 [1934]), 14–17.

Annalee Newitz | *My Laptop*

Vilém Flusser, "Digital Apparition," in *Electronic Culture: Technology and Visual Representation,* ed. Timothy Druckrey (New York: Aperture, 1996), 243–244.

Gail Wight | *Blue Cheer*

Michel Foucault, *Madness and Civilization: A History of Insanity in the Age of Reason,* trans. Richard Howard (New York: Random House, 1965), 197.

Julian Beinart | *The Radio*

Jean Baudrillard, "Design and Environment or How Political Economy Escalates into Cyberblitz," in *For a Critique of the Political Economy of the Sign,* trans. Charles Levin (St. Louis: Telos Press, 1981), 188–189.

Irene Castle McLaughlin | *The Bracelet*

Marcel Mauss, *The Gift: The Form and Reason for Exchange in Archaic Societies,* trans. W. D. Halls (New York: W. W. Norton, 2000 [1950]), 11–12, 14.

David Mitten | *The Axe Head*

Bruno Latour, *We Have Never Been Modern,* trans. Catherine Porter (Cambridge, Mass.: Harvard University Press, 1993), 67.

Susan Spilecki | Dit Da Jow

William James, *The Varieties of Religious Experience* (New York: Signet, 2003 [1902]), 49–50.

Nathan Greenslit | *The Vacuum Cleaner*

Karl Marx, *Capital: A Critique of Political Economy,* trans. Ben Fowkes (London: Penguin, 1976 [1867]), vol. I, 163.

William J. Mitchell | *The Melbourne Train*

Roland Barthes, *The Pleasure of the Text,* trans. Richard Miller (New York: Farrar, Straus and Giroux, 1975), 60–63, 66.

Judith Donath | *1964 Ford Falcon*

Igor Kopytoff, The "Cultural Biography of Things: Commoditization as Process," in *The Social Life of Things,* ed. Arjun Appadurai (Cambridge: Cambridge University Press, 1986), 66–68.

Trevor Pinch | *The Synthesizer*

Victor Turner, *The Forest of Symbols: Aspects of Ndembu Ritual* (Ithaca, N.Y.: Cornell University Press, 1967), 95, 98.

Tracy Gleason | *Murray: The Stuffed Bunny*

D. W. Winnicott, *Playing and Reality* (New York: Routledge, 1989 [1971]), 51.

David Mann | *The* World Book

Jacques Lacan, *Écrits: A Selection,* trans. Alan Sheridan (New York: W. W. Norton, 1977), 65, 68, 81.

Susan Rubin Suleiman | *The Silver Pin*

Melanie Klein, "Love, Guilt, and Reparation," in *Love, Guilt, and Reparation and Other Works, 1921–1945* (New York: Dell, 1975), 342–343.

Henry Jenkins | *Death-Defying Superheroes*

Umberto Eco, "The Myth of Superman," *Diacritics* 2, no. 1 (1972): 16.

Stefan Helmreich | *The SX-70 Instant Camera*

Susan Sontag, *On Photography* (New York: Dell, 1978), 8–9, 111.

Glorianna Davenport | *Salvaged Photographs*

Gaston Bachelard, *The Poetics of Space: The Classic Look at How We Experience Intimate Places,* trans. Maria Jolas (Boston: Beacon Press, 1994 [1958]), 5–6.

Susan Pollak | *The Rolling Pin*

Marcel Proust, *Remembrance of Things Past,* trans. C. K. Scott Montcrieff and Terence Kilmartin (New York: Vintage, 1981 [1913]), vol. I, 48–51.

Caroline A. Jones | *The Painting in the Attic*

Henri Bergson, *Matter and Memory,* trans. Nancy M. Paul and W. Scott Palmer (New York: Zone Books, 1990 [1896]), 133–135.

Olivia Dasté | *The Suitcase*

Sigmund Freud, "Mourning and Melancholia," in *The Standard Edition of the Complete Psychological Works of Sigmund Freud,* trans. and ed. James Strachey et al. (London: Hogarth Press, 1953–1974), vol. XIV, 249.

Nancy Rosenblum | *Chinese Scholars' Rocks*

Immanuel Kant, *The Critique of Judgment,* trans. James C. Meredith (Oxford: Oxford University Press, 1952 [1790]), 110–111.

Susannah Mandel | *Apples*

Lewis Thomas, *The Lives of a Cell: Notes of a Biology Watcher* (New York: Penguin, 1974), 3, 45.

Jeffrey Mifflin | *The Mummy*

Sigmund Freud, "The Uncanny," in *The Standard Edition of the Complete Psychological Works of Sigmund Freud,* trans. and ed. James Strachey et al. (London: Hogarth Press, 1953–1974), vol. XVII, 235.

Michael M. J. Fischer | *The Geoid*

Maurice Blanchot, *Awaiting Oblivion,* trans. John Gregg (Lincoln: University of Nebraska Press, 1997), 1–2.

Robert P. Crease | *Foucault's Pendulum*

Edmund Burke, *A Philosophical Enquiry into the Origin of Our Ideas of the Sublime and Beautiful* (Oxford: Oxford University Press, 1990 [1757]), 67.

Evelyn Fox Keller | *Slime Mold*

Mary Douglas, *Purity and Danger: An Analysis of Concepts of Pollution and Taboo* (New York: Routledge, 1992), 36.

Illustration Credits

Tod Machover | *My Cello*

Illustration on page 21 courtesy of the author.

Susan Yee | *The Archive*

Illustrations on pages 31 and 37 courtesy of AKG Images, Ltd.

Michelle Hlubinka | *The Datebook*

Illustrations on pages 77 and 85 courtesy of the author.

Gail Wight | *Blue Cheer*

Illustration on page 93 courtesy of Laura Splan. Illustrations on page 100 courtesy of the author.

Julian Beinart | *The Radio*

Illustrations on pages 103 and 109 courtesy of the author.

Irene Castle McLaughlin | *The Bracelet*

Illustrations on pages 111 and 117 courtesy of the author.

David Mitten | *The Axe Head*

Illustrations on pages 119 and 125 courtesy of R. Wing.

Note: All illustrations without specific credit are copyright-free stock photos, are public domain, or were created by the book's art director.

Susan Spilecki | Dit Da Jow

Illustration on page 127 courtesy of the author.

William J. Mitchell | *The Melbourne Train*

Illustration on page 145 courtesy of Ararat Railway Heritage.

Judith Donath | *1964 Ford Falcon*

Illustrations on pages 153 and 161 courtesy of the author.

Trevor Pinch | *The Synthesizer*

Illustrations on pages 163 and 169 courtesy of the author.

Tracy Gleason | *Murray*

Illustrations on pages 171 and 177 courtesy of the author.

Susan Rubin Suleiman | *The Silver Pin*

Illustrations on pages 185 and 192 courtesy of the author.

Henry Jenkins | *Death-Defying Superheroes*

Illustration on page 195 from "Detective Comics" #460 © 1976 DC Comics. All rights reserved. Used with permission. Illustration on page 206 from SPIDER-MAN: TM & © 2006 Marvel Characters, Inc. Used with permission.

Stefan Helmreich | *The SX-70 Instant Camera*

Illustration on page 209 courtesy of Oleg Volk of VOLK-STUDIO (www.olegvolk.net).

Glorianna Davenport | *Salvaged*

Illustrations on pages 217 and 223 courtesy of the author/Davenport archive.

Caroline A. Jones | *The Painting in the Attic*

Illustration on page 233 courtesy of the author.

Olivia Dasté | *The Suitcase*

Illustrations on pages 245 and 249 courtesy of GLOBE-TROTTER Ltd.

Nancy Rosenblum | *Scholars' Rocks*

Illustrations on pages 253 and 259 courtesy of the author/Rosenblum Family Collection.

Jeffrey Mifflin | *The Mummy*

Illustrations on pages 271 and 277 courtesy of the author.

Michael M. J. Fischer | *The Geoid*

Illustration on page 279 courtesy of K. H. Ilk, University Bonn. Illustration on page 285 courtesy of the author.

Robert P. Crease | *Foucault's Pendulum*

Illustrations on pages 287 and 295 courtesy of the Museum of Science, Boston.

Evelyn Fox Keller | *Slime Mold*

Illustration on page 306 courtesy of M. J. Grimson and R. L. Blanton, Biological Sciences Electron Microscopy Laboratory, Texas Tech University.

Index

"Affiliative Objects" (Suchman), 336n

Andersen, Hans Christian, 54

Appadurai, Arjun, 329n

Apple, the, 261–268, 307

Aramis or the Love of Technology (Latour), 339n

Archive, the, 31–36, 324

  digital, 323, 324

  physical, 32–35, 324

Archive Fever: A Freudian Impression (Derrida), 30, 341n

Art of the Natural World: Resonance of Wild Nature in Chinese Sculptural Art (Rosenblum and Doran), 333n

"Auguries of Innocence" (Blake), 333n

Autobiography of Benjamin Franklin, The, 330n

Awaiting Oblivion (Blanchot), 278

Axe Head, the, 119–125, 313

Bachelard, Gaston, 216, 342n

Bailey, Irma, 114–116

Ballet slippers, 55–60, 311

Barad, Karen, 336n

Barthes, Roland, 144, 316, 338n, 340n

Baudrillard, Jean, 8, 102, 139, 312, 331n, 338n

Beethoven, Ludwig van, 17

Beinart, Julian, 108, 312

Belmonte, Matthew, 75

Bergson, Henri, 232, 341n

Biography of objects, 139, 315. *See also* Kopytoff, Igor

Birth of the Clinic: An Archeology of Medical Perception, The (Foucault), 337n

Blake, William, 12, 254, 333n

Blanchot, Maurice, 278

Bleier, Ruth, 328n

"Blue Cheer," 93–100, 311

Bodies 2, 307

Bodies That Matter: On the Discursive Limits of "Sex" (Butler), 338n

Bracelet, the, 111–117

Bricolage (bricoleur), 4, 5, 11, 22, 309. See also Lévi-Strauss, Claude

engineer as bricoleur, 308 (see also Lévi-Strauss, Claude)

Brown, Bill, 329n

Budapest Diary (Suleiman), 188

Burgess, Anthony, 266, 333n

Burke, Edmund, 8, 286, 292

Butler, Judith, 338n

Capital: A Critique of Political Economy (Marx), 136, 331n, 338n

Causal explanations, preference for, 323

Cello, the, 13–20

Century of the Gene, The (Keller), 335n

Cevetello, Joseph, 68, 325–326

Childhood and Society (Erikson), 12

Chinese Scholars' Rocks, 8, 253–259, 319, 320

Clocks, 76, 310

Clockwork Orange, A (Burgess), 266, 333n

Clockwork Orange: A Play with Music, A (Burgess), 266, 333n

"Close-to-the-object" (as style of doing science), 300, 301

Clynes, Manfred, 68, 330n

Comics and collecting, 199, 201–205

Commodities, 312

  animation of, 139

Computer as creator of evocative objects, 43

Computer as evocative object, 43

Constructionism: Research Reports and Essays, 1985–1990 by the MIT Epistemology and Learning Group, The MIT Media Laboratory (Harel and Papert), 337n

Crease, Robert P., 294

Critique of Judgment (Kant), 252, 341n

Csikszentmihalyi, Mihaly, 329n, 341n

"Cultural Biography of Things: Commoditization as Process" (Kopytoff), 152

Cyborg, 68

Cyborg couplings, 325–326

"Cyborg Manifesto" (Haraway), 62

"Cyborgs and Space" (Clynes and Kline), 330n

Dasté, Olivia, 249, 317

Daston, Lorraine, 329n

Datebook, the, 77–84, 310–311

Davenport, Glorianna, 222

"DDS: Dynamics of Developmental Systems" (Keller), 335n

Death-defying superheroes, 7, 195–206

Derrida, Jacques, 8, 30, 324, 342n

"Design and Environment or How Political Economy Escalates into Cyber-Blitz" (Baudrillard), 102, 331n

Diabetes (monitoring, and "tight control" of, with objects), 65–68

Didion, Joan, 323

"Digital Apparition" (Flusser), 86

Digital database, 34

Disciplinary Society, 310–311

Discipline and Punish: The Birth of the Prison (Foucault), 337n

Discovering Reality: Feminist Perspectives on Epistemology, Metaphysics, Methodology, and Philosophy of Science (Harding and Hintikka, eds.), 328n–329n

Dit Da Jow, 127–134

Donath, Judith, 160, 314–315

Doran, Valerie, 333n

Douglas, Mary, 8, 296, 321, 342n

Durban, South Africa, 105

Eco, Umberto, 7, 194, 196, 197, 292–294, 332n, 334n

Écrits: A Selection (Lacan), 178, 339n, 340n

Education of Man (Froebel), 330n

"Elucidating Styles of Thinking through Learning about Knots" (Strohecker), 330n

Emergence, idea of, 42, 299

Epistemic Cultures: How the Sciences Make Knowledge (Knorr Cetina), 329n

Erasure (as used by Derrida), 324

Erikson, Erik, 12

"Essay on Criticism, An" (Pope), 332n

Evocative objects, notion of, 5, 10, 307

    activity of, 8 (*see also* Baudrillard, Jean; Latour, Bruno; Lévi-Strauss, Claude; Marx, Karl; Piaget, Jean; Vygotsky, Lev) as boundary objects, 8, 73, 307, 321, 322 (*see also* Douglas, Mary; Turner, Victor)

bringing philosophy down to earth, 8, 307 (see also Chinese Scholars' Rocks; Turkle, Sherry)

as catalysts of identity, 8, 156 (*see also* Erikson, Erik; Kopytoff, Igor)

computer as evocative object, 43 (*see also* Newitz, Annalee)

  as creator of evocative objects, 43 (*see also* Resnick, Mitchel)

holding power of, general, 8

identification of self with object (e.g., car), 156, 159

in integration of intellect and emotion, 5, 8, 309

as liminal ("threshold") objects, 8, 307, 315, 319, 320, 321 (*see also* Turner, Victor)

as transitional objects, 9, 314, 318 (*see also* Transitional objects; Winnicott, D. W.)

uncanny objects, as 8, 307, 322 (*see also* Freud, Sigmund)

"Fan" life and culture, 196–202

"Fate of the Transitional Object, The" (Winnicott), 332n

Feeling for the Organism: The Life and Work of Barbara McClintock, A (Keller), 328n, 335n

Feminist Approaches to Science (Bleier, ed.), 328n

"Fetishism" (Freud), 329n

Feynman, Richard, 328n

Fischer, Irene, 334n

Fischer, Michael M. J., 284

Flusser, Vilém, 86

Following the Equator (Twain), 146, 331n

For a Critique of the Political Economy of the Sign (Baudrillard), 338n

Ford Falcon, 7, 153–161, 314–315

Forest of Symbols: Aspects of Ndembu Ritual, The (Turner), 162

Foucault, Jean-Bernard-Leon, 289, 290

Foucault, Michel, 92, 310–311, 337n

Foucault pendulum, the, 287–295

Foucault's Pendulum (Eco), 292–294, 334n

Franklin, Benjamin, 331n

  daily planner, 78, 81

Franklin Institute (Philadelphia), 288, 289, 294

Freud, Sigmund
  on the after-effects of early bonding and sudden separation, 323
  on the fate of lost objects, 8, 317
  on fetishism, 329n
  Lacan's reading of, 319–320
  on "mourning," 8, 244, 318, 329n, 340n
  on the "uncanny," 8, 142, 270, 320, 329n, 342n

Froebel, Friedrich, 43, 330n

From Art to Science (Smith), 331n

"Function and Field of Speech and Language in
      Psychoanalysis, The" (Lacan), 340n

Furphy, Joseph, 150

Gardner, Howard, 51, 325

Garfinkel, Alan, 300, 301

"Gears of My Childhood" (Papert), 330n

Genetic Epistemology (Piaget), 38

*Geodesy? What's That? My Personal Involvement in the
      Age-Old Quest for the Size and Shape of the Earth*
      (I. Fischer), 280, 333n

Geoid, the, 279–284

*Gift: The Form and Reason for Exchange in Archaic Societies,
      The* (Mauss), 110, 338n

Gifts, 312. *See also* Mauss, Marcel

Gilligan, Carol, 328n

Gleason, Tracy, 176, 313, 314

Glucometer, the Elite, 6, 8, 63–68, 325–326

GNU/Emacs, 91

Greenslit, Nathan, 142

Growth of Logical Thinking from Childhood to Adolescence, The
      (Piaget and Inhelder), 328n

Haraway, Donna, 8, 62, 325–326, 330n, 336n, 343n

Harding, Sandra, 328n, 329n

Harel, Idit, 337n

Helmreich, Stefan, 215

Hintikka, Merrill B., 328n, 329n

Hlubinka, Michelle, 84, 310–311

Hung, Wong Fei, 128
Hyperinstruments, 19

Identification, child with object (knot), 27, 28
Identification with others through identification with object, 33
*In a Different Voice: Psychological Theory and Women's
    Development* (Gilligan), 328n
Inhelder, Barbel, 328n
*Insisting on the Impossible: The Life of Edwin Land, Inventor of
    Instant Photography* (McElheny), 332n
Instrumental vs. subjective technology (Turkle), 34
IRCAM (Boulez), 18

James, William, 9, 126
Jenkins, Henry, 206
Jones, Caroline A., 242
Joslin Diabetes Center, 65
Juilliard, 17, 18

Kant, Immanuel, 8, 252, 292, 294, 341n
Keller, Evelyn Fox, 306, 321–322, 328n, 335n, 342n
Keyboards, 17 51
Klein, Melanie, 184
Kline, Nathan, 330n
Knorr Cetina, Karin, 328n, 329n, 336n
Knot, the, 23–29
Knot Lab, 25–29, 309–310
Knots, 8, 309
Kopytoff, Igor, 8, 152, 315
Kristeva, Julia, 70, 338n
Kung fu, 128–132

Laboratory Life: The Construction of Social Scientific Facts
    (Latour and Woolgar), 328n, 339n
Lacan, Jacques, 178, 316–317, 340n
Lam, Wing, 129
Language
  as constitutive of self, 178, 317
  as liminal object, 316, 317
Laptop (computer), 8, 87–91, 325, 325–326

La Sylphide (ballet), 56

Latour, Bruno, 118, 313, 328n, 329n, 339n

Lawson, Henry, 148, 149, 332n

Le Corbusier (Charles-Edouard Jeanneret), 32, 33, 324

Lego Group, 44

"Letters to a Young Poet" (Rilke), 330n

Lévi-Strauss, Claude, 4, 6–7, 22, 308, 309, 328n, 336n

Life on the Screen: Identity in the Age of the Internet, The (Turkle), 328n

Liminal ("threshold") objects, 8, 307, 315, 316, 319, 320

Linux, 91

Lives of a Cell: Notes of a Biology Watcher, The (Thomas), 260, 333n

Logo turtle, 43, 44

"Love, Guilt, and Reparation" (Klein), 184

Ludiomil, 94, 95, 98, 99

Machover, Tod, 20

Madeleine (cookie), 226, 227, 228, 318

Madness and Civilization: A History of Insanity in the Age of Reason (Foucault), 92, 337n

Magritte, René, 267

Mandel, Susannah, 268

Mangle of Practice: Time, Agency, and Science, The (Pickering), 336n

Mann, David, 183, 316, 316–317

Manufacture of Knowledge: An Essay on the Constructivist and Contextual Nature of Science, The (Knorr Cetina), 328n

Massachusetts General Hospital (Boston), 272

Mastery, play as strategy for achievement of (Erikson), 12

Matter and Memory (Bergson), 232, 341n

Marx, Karl, 8, 136, 139, 312, 338n

Mauss, Marcel, 110, 312, 338n. *See also* Gifts

McClintock, Barbara, 300, 301, 322

McElheny, Victor, 332n

McLaughlin, Irene Castle, 117

Meaning of Things: Domestic Symbols and the Self, The (Csikszentmihalyi and Rochberg-Halton), 329n, 341n

Medina, Eden, 60, 311

Melbourne Train, the, 7, 145–151, 315, 316

Memory closet, 3, 4, 10

Meritol, 98

Mifflin, Jeffrey, 276, 320–321

Mind in Society (Vygotsky), 46

Mindstorms: Children, Computers, and Powerful Ideas
    (Papert), 309, 337n

Mitchell, William J., 150, 315, 316

MIT Media Lab, 19, 24

Mitten, David, 125, 313

Moog synthesizer, 164

"Mourning and Melancholia" (Freud), 244, 329n, 340n

"Mr. B." (case study), 229, 230, 318 319

Mumford, Lewis, 76, 310, 337n

Mummy, the ("Padihershef"), 271–277, 320, 321

"Myth of Superman" (Eco), 194, 332n

Mythologies (Barthes), 337n

Nanney, David L., 301, 335n

Neumann, John von, 305

"Never-Never Land, The" (Lawson), 332n

Newitz, Annalee, 91, 325, 325

Objects, examples of actions of
  as anthropomorphic beings, 138–140
  in body discipline, 311
  to bring society within the self, 311
  as commodities, 136, 139, 312
  to internalize sense of time, 310
  within mourning process, 318
  to reveal relationships to nature, history, etc., 312
  to reveal social relations, 312
Objects of science as objects of passion, 8, 74, 212, 309
Object (or "thing") studies, 5
Objects as sublime, 286, 252
Objects as symbols of immigrant status, 186–192
On Photography (Sontag), 208
"On a Question Preliminary to Any Possible Treatment of
    Psychosis" (Lacan), 340n
"On the Signification of the Phallus" (Lacan), 339n

Painting in the attic, the, 233–242

Papert, Seymour, 43, 309, 330n, 337n

  gears of, 43, 309

Paradoxes, 40, 41, 43

Pascal, Blaise, 276, 333n

Pasteurization of France (Latour), 338n, 339n

Paterson (Williams), 336n

Pensées (Pascal), 333n

Philosophical Enquiry into the Origin of Our Ideas of the Sublime and Beautiful, A (Burke), 286

Photography as ritual of family life, 208

Piaget, Jean, 6, 38, 308, 309, 323, 328n

Pickering, Andrew, 336n

Pinch, Trevor, 168, 325

Playing and Reality (Winnicott), 170, 309, 339, 332n, 337n, 339n

Pleasure of the Text, The (Barthes), 144, 339n

Poetics of Space: The Classic Look at How We Experience Intimate Places (Bachelard), 216, 341n

Pollak, Susan, 231, 318–319

Pollution and cleanliness, 321. *See also* Douglas, Mary

Pope, Alexander, 150, 332n

"Posthumanist Performativity: Toward an Understanding of How Matter Comes to Matter" (Barad), 336n

"Processes and Products for Forming Photographic Images in Color" (Rogers), 332n

Programmable bricks, 44

Property, Substance, and Effect: Anthropological Essays on Persons and Things (Strathern), 336n

Proust, Marcel, 224, 318, 319, 332n, 341n

Psychoanalytic Politics: Jacques Lacan and Freud's French Revolution (Turkle), 337n

Psychopharmaceuticals and identity, 311

Pull-toy, 308, 309

Pull-toy paradox (Resnick), 40, 41, 42, 308

Purity and Danger: An Analysis of the Concepts of Pollution and Taboo (Douglas), 296, 321, 341n

Pyle, Howard, 263

Radio (mute), 103–109, 312

Ratey, John J., 90, 331n

Red Shoes, The (Andersen), 54

Reflections on Gender and Science (Keller), 335n, 341n

Remembrance of Things Past (Proust), 224, 227, 228, 230, 319, 332n, 341n

Resnick, Mitchel, 44, 308, 309, 323, 330n, 337n

Rilke, Rainer Maria, 68, 330n

Rites of Passage (van Gennep), 339n

Rites of passage, 315, 319. *See also* Liminal objects; Turner, Victor; Van Gennep, Arnold

Ritual Process: Structure and Anti-Structure, The (Turner), 329n, 339n

Rochberg-Halton, Eugene, 329n, 341n

Rogers, Howard G., 210, 332n

"Role of the Cytoplasm in Heredity" (Nanney), 335n

Rolling pin, the, 225–231, 307, 318

Rosenblum, Anna, 257

Rosenblum, Nancy, 258, 319

Rosenblum, Richard, 257, 258, 333n

Salvaged photographs, 217–222

Samaras, Lucas, 210, 213

Savage Mind, The (Lévi-Strauss), 22, 328n, 336n

Science in Action: How to Follow Scientists and Engineers through Society (Latour), 339n

Second Self: Computers and the Human Spirit (Turkle), 330n

Segel, Lee, 298, 300, 301, 322

Selye, Hans, 331n

Shortwave radio, 166

Silver pin, the, 7, 185–192

Simians, Cyborgs, and Women: The Reinvention of Nature (Haraway), 336n

Slime mold, 297–306, 321–322
  as boundary object, 321

Smith, Cyril Stanley, 331n

"Sociality with Objects: Social Relations in Postsocial Knowledge Societies" (Knorr Cetina), 336n

Social Life of Things: Commodities in Cultural Perspective, The (Appadurai, ed.), 329n

Sontag, Susan, 9, 208

Spilecki, Susan, 134

Stallman, Richard, 90

StarLogo, 44, 323
Stars, 39–45, 308
Strathern, Marilyn, 336n
Strangers to Ourselves (Kristeva), 70, 338n
Stress, 138, 139
"Stress and Psychiatry" (Selye), 331n
Strohecker, Carol, 29, 309, 310, 330n
Stuffed bunny, the (Murray), 171–176, 313, 314
Suchman, Lucy, 336n
Suitcase, the, 245–249, 317
Suleiman, Susan Rubin, 192
Surely You're Joking, Mr. Feynman (Feynman), 328n
SX-70 Instant Camera, 209–215, 307
Synthesizer, the, 163–168
System of Objects, The (Baudrillard), 139, 331n

Taglioni, Marie, 56
Technics and Civilization (Mumford), 76, 337n
Tenner, Edward, 329n
Things (Brown, ed.), 329n
Things That Talk: Object Lessons from Art and Science
     (Daston, ed.), 329n
Thomas, Lewis, 9, 260, 268, 333n
Torvalds, Linus, 91
Transitional object, 8, 228, 241, 242, 313, 314, 315, 319
  Murray, the stuffed bunny (as example of), 172–176
Turkle, Sherry, 3–10, 43, 307–326, 328n, 330n, 337n
Turner, Victor, 8, 162, 315, 319, 321, 329n, 339n
*Turtles, Termites, and Traffic Jams: Explorations in Massively
     Parallel Microworlds* (Resnick), 330n, 337n
Twain, Mark, 146, 332n

"Uncanny, The" (Freud), 142, 270, 329n, 341n
"Understanding Topological Relationships through
     Comparisons of Similar Knots" (Strohecker), 330n
*User's Guide to the Brain: Perception, Attention, and the Four
     Theaters of the Brain, A* (Ratey), 331n

Vacuum cleaner, the, 137–142
Van Gennep, Arnold, 339n

Varieties of Religious Experience, The (James), 126
VCS-3 electronic music synthesizer, 164, 165
Violins, 14, 15, 19
Virtual vs. physical objects, 323, 324
Vygotsky, Lev, 46

Wah, Tan Kwok, 128
Web site for car, 157–159
We Have Never Been Modern (Latour), 118, 329n, 339n
Western Native Township, Johannesburg, 107, 109
When Things Bite Back (Tenner), 329n
Whitman, Walt, 9
"Why Knot?" (Strohecker), 330n
Wight, Gail, 100, 311
Williams, William Carlos, 308, 336n
Winnicott, D. W., 8, 170, 228, 241, 242, 309, 313, 314, 315,
    319, 333n, 337n, 339n. *See also Playing and Reality;*
    Transitional objects
Wireless World Synthesizer, 166, 167
WizNet (multi-user chat system), 88, 89
Woolgar, Steven, 328n, 339n
"Word Processor, The" (Derrida), 342n
World Book (Encyclopedia), 179–183, 316
  as means of access to language, 178–181

Year of Magical Thinking, The (Didion), 317–318, 340n
Yee, Susan, 36, 324–325
Yellow raincoat, the, 71–75